Cosmic Ray Anisotropy Workshop

(CRA 2013)

Journal of Physics: Conference Series Volume 531

Madison, Wisconsin, USA
26 – 28 September 2013

ISBN: 978-1-63439-312-6
ISSN: 1742-6588

Printed from e-media with permission by:

Curran Associates, Inc.
57 Morehouse Lane
Red Hook, NY 12571

Some format issues inherent in the e-media version may also appear in this print version.

Printed by Curran Associates, Inc. (2014)

Institute of Physics
Dirac House, Temple Back
Bristol BS1 6BE UK

Phone: 44 1 17 929 7481
Fax: 44 1 17 920 0979

techtracking@iop.org

Additional copies of this publication are available from:

Curran Associates, Inc.
57 Morehouse Lane
Red Hook, NY 12571 USA
Phone: 845-758-0400
Fax: 845-758-2634
Email: curran@proceedings.com
Web: www.proceedings.com

TABLE OF CONTENTS

Journal of Physics: Conference Series **531** (2014) 011001 doi:10.1088/1742-6596/531/1/011001

Preface

In the search for the origin of cosmic rays in our galaxy, various observations are combined as pieces of a complex puzzle to account for the variability of the galactic sources and their local environments, as well as of the properties of the interstellar medium in which cosmic particles propagate. In this puzzle, multi-wavelength observations aim to pinpoint the properties of cosmic-ray sources, based on electromagnetic and neutrino emissions associated with hadronic acceleration processes. The detailed study of the energy spectrum and composition of cosmic rays on Earth, on the other hand, aims to probe the combined and interwoven effects of injection, by acceleration processes, and propagation in the interstellar medium. The observation of arrival directions of cosmic rays on Earth, in addition, potentially provides valuable information on the distribution of the closest and more recent active galactic sources as well as on the properties of the local interstellar magnetic field.

The quasi-isotropic distribution of galactic cosmic rays tells the global story of their journey from their sources. The turbulent magnetized galactic medium sufficiently scrambles the arrival directions of cosmic rays on Earth so that their main injection direction is concealed. On the other hand, the observations of a small but significant energy-dependent anisotropy are starting to provide clues on how cosmic rays propagate throughout the local interstellar medium. Anisotropy observations have been used to study diffusion properties of GeV cosmic rays inside the termination shock of the heliosphere and their dependencies on solar cycles. At TeV energy, cosmic rays are sensitive to larger-scale magnetic structures, such as the turbulent boundary between the heliosphere and the local interstellar medium or its elongated tail. At higher energy, it is expected that the interstellar magnetic field within the particle mean free path has major contributions to their arrival directions on Earth.

In this workshop, we addressed the potential use of high-energy cosmic-ray anisotropy observations as a probe into the properties of particle transport in astrophysical magnetized plasmas, such as the heliosphere and the local interstellar medium. Along with experts from the fields of astronomy, astrophysics, plasma physics, heliospheric physics and interstellar medium, we discussed how each field can contribute to the understanding of cosmic-ray propagation in our local interstellar magnetic field. This will represent another piece in the search for cosmic-ray sources in the galaxy.

Peer review statement

All papers published in this volume of *Journal of Physics: Conference Series* have been peer reviewed through processes administered by the proceedings Editors. Reviews were conducted by expert referees to the professional and scientific standards expected of a proceedings journal published by IOP Publishing.

Journal of Physics: Conference Series **531** (2014) 012001 doi:10.1088/1742-6596/531/1/012001

Studies of the cosmic ray spectrum and large scale anisotropies with the KASCADE-Grande experiment

A. Chiavassa[1], W.D. Apel[2], J.C. Arteaga-Velázquez[3], K. Bekk[2], M. Bertaina[1], J. Blümer[2,4], H. Bozdog[2], I.M. Brancus[5], E. Cantoni[1,6,13], F. Cossavella[4,14], C. Curcio[1], K. Daumiller[2], V. de Souza[7], F. Di Pierro[1], P. Doll[2], R. Engel[2], J. Engler[2], B. Fuchs[4], D. Fuhrmann[8,15], H.J. Gils[2], R. Glasstetter[8], C. Grupen[9], A. Haungs[2], D. Heck[2], J.R. Hörandel[10], D. Huber[4], T. Huege[2], K.-H. Kampert[8], D. Kang[4], H.O. Klages[2], K. Link[4], P. Łuczak[11], M. Ludwig[4], H.J. Mathes[2], H.J. Mayer[2], M. Melissas[4], J. Milke[2], B. Mitrica[5], C. Morello[6], J. Oehlschläger[2], S. Ostapchenko[2,16], N. Palmieri[4], M. Petcu[5], T. Pierog[2], H. Rebel[2], M. Roth[2], H. Schieler[2], S. Schoo[2], F.G. Schröder[2], O. Sima[12], G. Toma[5], G.C. Trinchero[6], H. Ulrich[2], A. Weindl[2], J. Wochele[2], J. Zabierowski[11]

[1] Dipartimento di Fisica, Università degli Studi di Torino, Italy
[2] Institut für Kernphysik, KIT - Karlsruher Institut für Technologie, Germany
[3] Universidad Michoacana, Instituto de Física y Matemáticas, Morelia, Mexico
[4] Institut für Experimentelle Kernphysik, KIT - Karlsruher Institut für Technologie, Germany
[5] National Institute of Physics and Nuclear Engineering, Bucharest, Romania
[6] Osservatorio Astrofisico di Torino, INAF Torino, Italy
[7] Universidade São Paulo, Instituto de Física de São Carlos, Brasil
[8] Fachbereich Physik, Universität Wuppertal, Germany
[9] Department of Physics, Siegen University, Germany
[10] Dept. of Astrophysics, Radboud University Nijmegen, The Netherlands
[11] National Centre for Nuclear Research, Department of Cosmic Ray Physics, Lodz, Poland
[12] Department of Physics, University of Bucharest, Bucharest, Romania

E-mail: andrea.chiavassa@to.infn.it

Abstract. KASCADE-Grande is an air shower observatory devoted to the detection of cosmic rays in the $10^{16} - 10^{18} eV$ energy range. For each event the arrival direction, the total number of charged particles (N_{ch}) and the total number of muons (N_{μ}), at detection level (i.e. 110 m a.s.l.), are measured. The detection of these observarbles, with high accuracy, allows the study of the primary spectrum, chemical composition and large scale anisotropies, that are the relevant informations to investigate the astrophysics of cosmic rays in this energy range. These studies are of main importance to deeply investigate the change of slope of the primary spectrum detected at $\sim 4 \times 10^{15} eV$, also known as the knee, and to search for the transition from galactic to extra-galactic cosmic rays.

[13] now at: Istituto Nazionale di Ricerca Metrologia, INRIM, Torino.
[14] now at: DLR Oberpfaffenhofen, Germany.
[15] now at: University of Duisburg-Essen, Duisburg, Germany.
[16] now at: University of Trondheim, Norway.

The two-dimensional (N_{ch} vs N_μ) spectrum is the basis for the cosmic-ray chemical composition studies. EAS and detection fluctuations prevent the measurement of the primary mass on an event by event basis, nevertheless the precision obtained by the KASCADE-Grande experiment allows to separate events into mass groups. A search for anisotropies in the arrival directions of primary cosmic rays has been performed using the East-West analysis technique with the events detected by the Grande array. First harmonic modulation has been derived from the distributions (in sidereal, solar and anti-sidereal times) of the counts difference from the East and West sectors.

In this contribution we present and discuss the latest experimental results.

1. Introduction

To unveil the origin, nature, propagation and acceleration mechanism of galactic cosmic rays, high resolution measurements with high statistics of the energy spectrum, chemical composition and arrival directions are needed. Due to the low fluxes, at high energies ($E > 1PeV$), cosmic rays must be studied by means of extensive air showers (EAS) experiments and the characteristics of the primary particles are inferred indirectly.

The energy of the primary cosmic ray that originated the EAS can be obtained either from the total number of charged particles or from the muon number or from a combination of these two observables. In both cases the absolute energy scale depends on the hadronic interaction model used in the EAS development simulation.

The primary chemical composition studies are based on the detection of different EAS parameters[1]: the atmospheric depth of the shower maximum (available to fluorescence light detectors), the correlation between the muon and electron numbers at observation level (that can be detected in various ways, mainly by scintillation counters). The interpretation of these data requires full EAS simulations that are necessarily based on hadronic interaction models founded on the extrapolation of the accelerator measurements, performed at lower energies. Data of the LHC experiments will cover the energies of the knee, thus in the near future (tuning the hadronic interaction models with these data) we can expect great improvements of the situation. Nevertheless shower development fluctuations almost prevent an event by event detection of the nature of the primary particle.

The measured and expected amplitude of large scale anisotropies, for energies below $10^{16}eV$, are at the level of $10^{-4} \div 10^{-3}$. The significance of the measurements depends on the number of events in the data set, so long duration data taking are required. At the same time the counting rate variations induced by atmospheric effects (i.e. pressure and temperature) are bigger than those expected by cosmic rays anisotropies and so the arrays stability plays a crucial role in such analysis. To take into account the instabilities induced by atmospheric and instrumental effects the analysis performed by the KASCADE-Grande collaboration uses the East-West technique[2].

2. The Experiment

The KASCADE-Grande detector (located at 49.1° N, 8.4° E, $110m a.s.l.$) was the successor of the KASCADE experiment[3] and incorporated the original electromagnetic detectors and muon devices of KASCADE to a bigger system of detectors, called Grande, composed of a $700 \times 700m^2$ array with $37 \times 10m^2$ scintillator stations regularly spaced by an average distance of $137m$[4]. A smaller array, named Piccolo, was also added and was used to coordinate the KASCADE and Grande triggers. The layout of the experiment is presented in figure 1.

The Grande array was used to sample the density of charged particles of the shower front at ground level and to measure the particle arrival times of the EAS. The core position, the number of charged particles (N_{ch}) and the arrival direction of the shower were extracted from the Grande data through an iterative fit and a careful modeling of the EAS front. To reconstruct

Journal of Physics: Conference Series **531** (2014) 012001 doi:10.1088/1742-6596/531/1/012001

the arrival direction of the shower from the arrival times a fit was applied assuming a curved shower front as suggested by CORSIKA/QGSJET II simulations. On the other hand, to obtain the core position and the shower size from the density of charged particles of the EAS at ground a modified NKG lateral distribution function was fitted by means of a maximum-likelihood procedure[4].

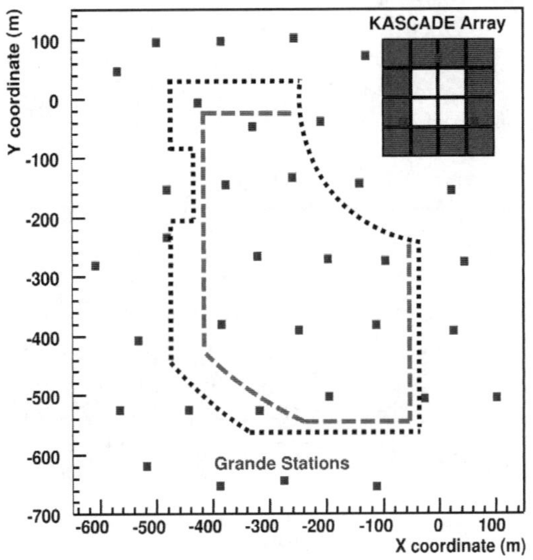

Figure 1. Layout of the KASCADE-Grande experiment. The KASCADE array and the distribution of the 37 stations of the Grande array are shown. The 192 muon detectors are placed in the outer 12 clusters of the KASCADE array (hatched area). The dashed line shows the fiducial area selected for the all-particle and heavy mass group spectra analysis. The dotted line indicates the area used for the measurement of the light mass group spectrum.

An important component of this experiment was the KASCADE muon array, composed by $192 \times 3.2m^2$ shielded scintillator detectors, which are sensitive to muons with threshold energy above $230MeV$ for vertical incidence. With the Grande information and the measurements of the shielded array of the lateral distribution of muons in the shower front, the muon size (N_μ) was reconstructed event-by-event at KASCADE-Grande. The procedure involved a maximum loglikelihood fit along with a Lagutin-Raikin distribution function[4].

Buried under several layers of soil, sand, iron and concrete, close to the center of the KASCADE array, was located the tunnel of the muon tracking detector (MTD), which was composed of streamer tubes grouped in modules[5]. This system was used to reconstruct the individual tracks of penetrating particles of a section of the EAS and to measure the muon pseudorapidities. By applying a triangulation procedure on the reconstructed tracks the muon production height of each event was also derived[6].

Systematic uncertainties for the core, N_{ch} and arrival direction of the EAS are studied directly[4] by comparing the results of the Grande and KASCADE reconstructions, which work independently. Comparisons were performed for a subset of data with cores located inside a common area for both detectors and shower sizes in the interval $log_{10}N_{ch} = 5.8 \div 7.2$. Accuracies of the EAS core positions and arrival directions are found to be of the order of $\sim 5m$ and $\sim 0.7°$, respectively. Meanwhile, for the total number of charged particles a resolution $\leq 15\%$ is achieved. Those values are in full agreement with expectations from Monte Carlo simulations.

The resolution on the N_μ EAS parameter is evaluated reconstructing simulated events, a

Journal of Physics: Conference Series **531** (2014) 012001 doi:10.1088/1742-6596/531/1/012001

$\sim 20\%$ accuracy has been achieved. More details on the experimental setup and on the event reconstruction can be found in[4].

3. Results

3.1. All particle energy spectrum

The energy of the primary particle that originated the detected EAS is determined by the KASCADE-Grande experiment by means of the N_{ch} and N_μ observables[7], combining these two variables indeed we can lower the dependence from the chemical composition of the primary particles. This is obtained evaluating for each event the so called k parameter, that is essentially a measurement of the ratio between the muon and the charged particles numbers.

$$k = \frac{\log_{10}(N_{ch}/N_\mu) - \log_{10}(N_{ch}/N_\mu)_H}{\log_{10}(N_{ch}/N_\mu)_{Fe} - \log_{10}(N_{ch}/N_\mu)_H} \tag{1}$$

$$\log_{10}(N_{ch}/N_\mu)_{H,Fe} = c_{H,Fe}\log_{10} N_{ch} + d_{H,Fe} \tag{2}$$

From its definition is clear that k is a number centered around zero (one) for proton (iron) generated events, if expressed as a function of N_{ch} for Monte Carlo events, assuming intermediate values for all other primaries. The values of the k parameters are tuned by a full EAS and detector simulation, the analysis reported in[7] is based on the QGSJetII-02[8] hadronic interaction model. Having calculated, for each event, the k parameter the primary energy is estimated from the N_{ch} value:

$$\log_{10}(E/GeV) = [a_H + (a_{Fe} - a_H)\cdot k]\log_{10}(N_{ch}) + b_H + (b_{Fe} - b_H)\cdot k$$

To take into account the shower evolution in atmosphere the parameters, $a_{H,Fe}, b_{H,Fe}, c_{H,Fe}, d_{H,Fe}$, contained in the k and E expressions are derived in five different angular intervals, whose upper limits are: $16.7°, 24.0°, 29.9°, 35.1°$ and $40.0°$. The values of the parameters can be found in[7].

The all-particle energy spectrum is then measured in the five different angular bins. As shown in[7] these spectra are slightly shifted, indicating that the EAS evolution in atmosphere is not correctly described by the simulations. Nevertheless these differences are inside the experimental uncertainties and thus we mediate them to obtain the all particle energy spectrum measured in zenith angle range from 0° to 40°. The residuals of the all-particle energy spectrum multiplied by a factor, in such a way that the middle part of the spectrum becomes flat, are shown in figure 2.

The measured spectrum cannot be described by a single power law: a hardening around $10^{16}eV$ and a steepening at $\log_{10}(E/eV) = 16.92 \pm 0.10$ are observed. The statistical significance of the steepening is 2.1σ, here the change of the spectral slope is from $\gamma = -2.95 \pm 0.05$ to $\gamma = -3.24 \pm 0.08$. The same spectral features are meanwhile confirmed by the Tunka-133[9] and Ice-Top[10] experiments.

This procedure relies on the EAS simulation and thus depends on the high-energy hadronic interaction model used. To evaluate the systematic effects introduced in the all-particle spectrum measurement[11] the same procedure has been repeated using events simulated with the SIBYLL2.1[12], EPOS1.99[13] and QGSJetII-04[14] hadronic interaction models.

Applying the energy calibration functions, obtained by each model, to the measured data the all-particle energy spectra for the five zenith angle bins are obtained for the four previously mentioned models. Different sources of uncertainty affect the all-particle energy spectrum: the total systematic uncertainty is $\sim 20\%$ at the threshold ($E = 10^{16}eV$) and $\sim 30\%$ at the highest energies ($E = 10^{18}eV$) almost independently from the interaction model used to interpret the data. The final all-particle spectrum of KASCADE-Grande is obtained (see figure 3) by combining the spectra for the individual angular ranges. In general the shape of the energy

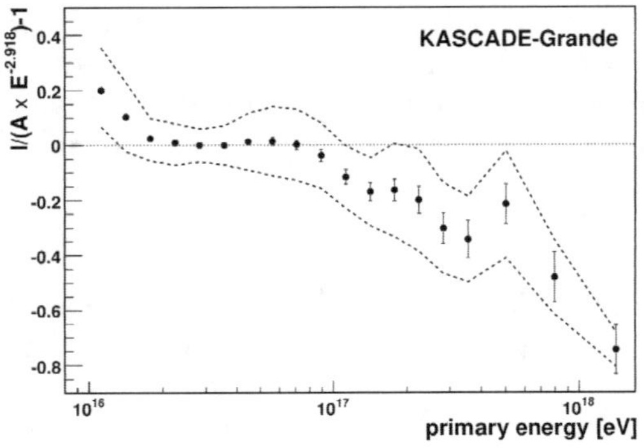

Figure 2. The all-particle energy spectrum obtained with KASCADE-Grande. The residual intensity after multiplying the spectrum with a factor of $E^{2.918}$ is displayed as well as the band of systematic uncertainty.

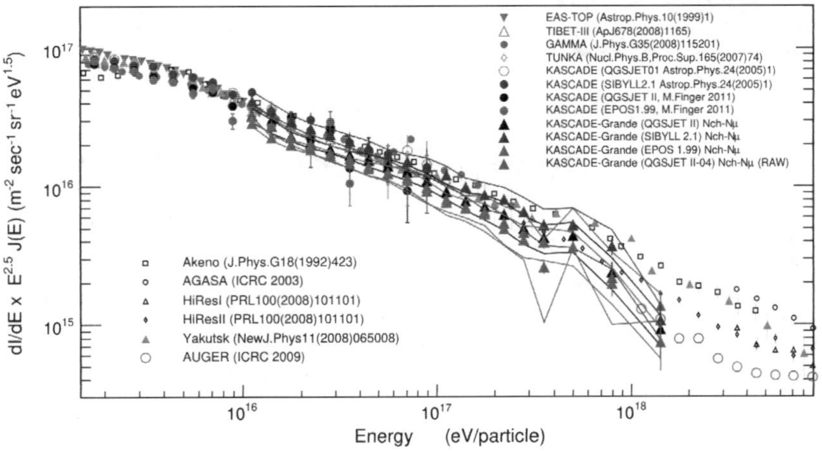

Figure 3. Comparison of the all-particle energy spectrum obtained with KASCADE-Grande data based on SIBYLL2.1 (blue), QGSJetII-02 (black), QGSJetII-04 (pink) and EPOS1.99 (red) models to results of other experiments. The band denotes the systematic uncertainties in the flux estimation

spectrum is very similar for all models, however, a shift in flux is clearly observed which amounts to $\sim 25\%$ increase in case of SIBYLL2.1 and $\sim 10\%$ decrease in case of EPOS1.99. The features seen in the spectrum are not an artifact of the hadronic interaction model used to interpret the data but they are in the measured data. In the overlapping region, KASCADE-Grande data are compatible inside the systematic uncertainties with KASCADE data interpreted with the same model.

3.2. Energy spectra of individual mass groups

The k parameter previously defined can also be used to separate the events in samples generated by two different primary mass groups. To emphasize the features of the heavy elements we selected the electron-poor events with $k_{ep}(E) \geq (k_C(E) + k_{Si}(E))/2$, i.e. events with a k value greater than the mean value of the expectations for C and Si primaries (QGSJetII-02[8] based simulation). The spectra of these event samples are shown in figure 4, the band indicates changes of the spectra when the cut is varied by one standard deviation in the $k_{ep}(E)$ definition.

The reconstructed spectrum of the electron-poor events shows a distinct knee like feature at about $8 \times 10^{16} eV$[16]. Applying a fit of two power laws to the spectrum interconnected by a smooth knee[15] results in a statistical significance of 3.5σ that the entire spectrum cannot be described with a single power-law. The change of the spectral index is $\Delta\gamma = -0.48 \pm 0.05$ from $\gamma = -2.76 \pm 0.02$ to $\gamma = -3.24 \pm 0.05$ with the break position at $\log_{10}(E/eV) = 16.92 \pm 0.04$. The spectrum of the electron-rich events (corresponding, with this cut definition, to light and medium mass primaries) is compatible with a single power law with slope index $\gamma = -3.18 \pm 0.01$. A recovery to a harder spectrum at energies greater than $10^{17} eV$ cannot be excluded by this analysis.

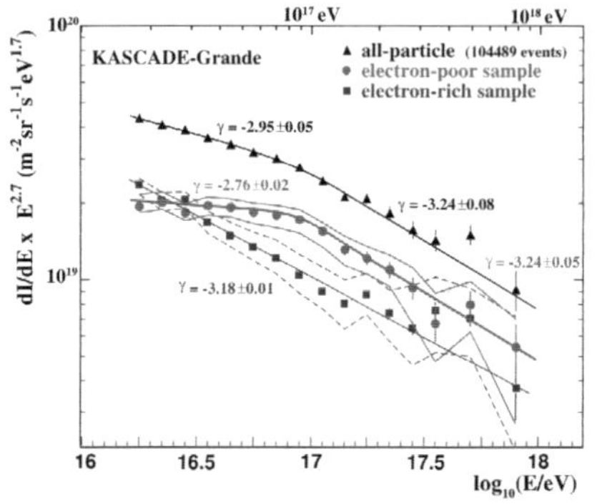

Figure 4. Reconstructed energy spectrum of the electron-poor and electron-rich components together with the all-particle spectrum for the angular range $0° - 40°$. The error bars show the statistical uncertainties; the bands assign systematic uncertainties due to selection of the subsamples.

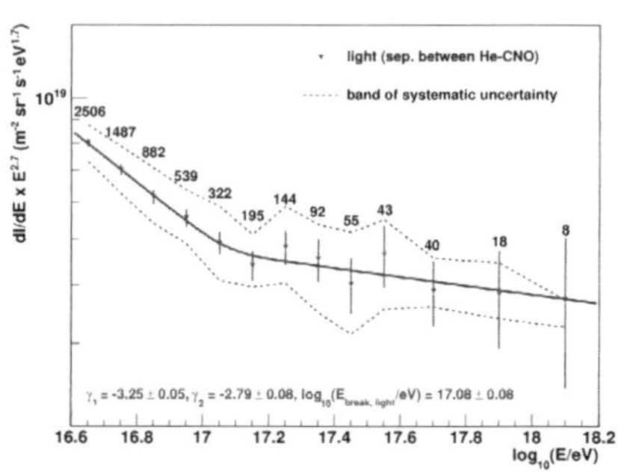

Figure 5. The reconstructed energy spectrum of the light mass component of cosmic rays. The number of events per energy bin is indicated as well as the range of systematic uncertainty. The error bars show the statistical uncertainties.

To increase the statistics and deeply investigate this possible hardening of the light primaries spectrum a larger fiducial area has been defined, essentially accepting events at larger distances from the muon detector (i.e. from the KASCADE array, see figure 1). The main effect of this event selection is that the 100% efficiency is reached at higher energies, that is not a problem for this analysis aimed to study a possible spectral feature at energies greater than $10^{17} eV$. In order to emphasize features of the light mass group we redefine the cut on the k parameter as $k_{er}(E) \leq (k_C(He) + k_C(E))/2$ (again a simulation based on the QGSJetII-02 hadronic interaction model is used). The obtained spectrum is shown in figure 5; a hardening, or ankle-like feature, is clearly observed[17]. Fitting this spectrum with the same function used for the all-particle and heavy mass groups primary spectra we obtain a change of the spectral index from $\gamma = -3.25 \pm 0.05$ to $\gamma = -2.79 \pm 0.08$ at an energy of $\log_{10}(E/eV) = 17.08 \pm 0.08$. The measured number of events above the bending is $N_{meas} = 595$. Without the bending we would expect $N_{exp} = 467$ events above this ankle-like feature. The Poisson probability to measure at least N_{meas} events above the bending, if N_{exp} are expected, is $P \sim 7.23 \times 10^{-9}$, corresponding to a 5.8σ significance.

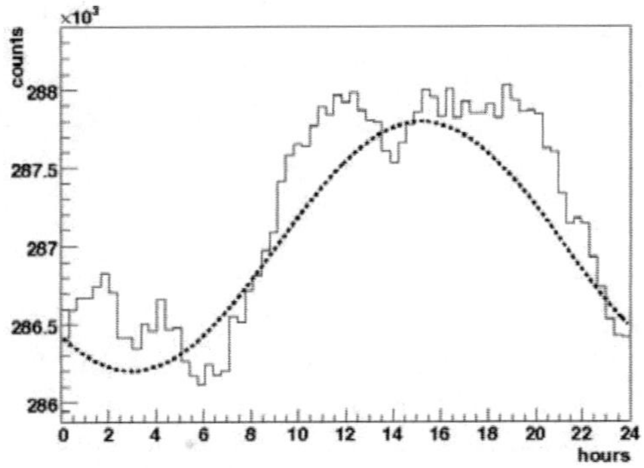

Figure 6. Sidereal time distribution of the number of counts ($\theta < 40°$ and $LogN_{ch} > 5.2$) in 20 minute intervals obtained applying the East-West method. The dashed line shows the calculated first harmonic.

Comparing the two previous observations it is important to notice that the knee in the heavy component occurs at a lower energy compared to the bending in the spectrum of the light primaries. Therefore the steepening of the heavy spectrum and the recovery of the light component are not due to a bias in the reconstruction or separation procedures. It is worth pointing out that the slope of the heavy mass spectrum above the knee-like feature is very similar to the slope of the light mass spectrum before the ankle-like feature. The slope index of the light mass group spectrum above the ankle-like feature is $\gamma \sim -2.7$ and this can be interpreted as an indication of an injection of a new (extragalactic) population of high energy cosmic-rays[17].

3.3. Large Scale Anisotropies

The search for large scale anisotropies has been performed through a differential method, the so called East-West[2] method, based on the counting rate differences between East-ward and West-ward directions. This method allows to remove counting rate variations caused by atmospheric and instrumental effects. The used data set contains 10^7 events recorded between December 2003 and October 2011. To ensure reconstruction quality, a cut on the zenith angle θ and on the number of charged particles at observation level (N_{ch}) was applied: $\theta < 40°$ and $Log(Nch) > 5.2$.

Figure 6 shows the modulation in sidereal[17] time obtained using the East-West method. The amplitude of the first harmonic calculated in sidereal time is $(0.28 \pm 0.08) \times 10^{-2}$ with a 0.2% Rayleigh probability of being due to background fluctuation (i.e. $\sigma = 3.5$) at median energy $3.3 \times 10^{15} eV$. The 99% C.L. upper limit on the amplitude is 0.47×10^{-2}, derived according to the distribution drawn from a population characterized by an anisotropy of unknown amplitude and phase as derived by Linsley[18].

To investigate a variation of the amplitude and phase of the first harmonic with primary energy we have performed the same analysis in intervals of the number of charged particles. The results are shown in table 1, in none of the bins the amplitude of the harmonic is statistically significant and so we have calculated the upper limits at 99% confidence level. The N_{ch} limits of the intervals used for the harmonic analysis are converted to primary energy and the median energy of the events in each bin is defined as representative energy.

[17] sidereal time: Common time scale among astronomers which is based on the Earth's rotation measured relative to the fixed stars.

$\log_{10}(N_{ch})$	E (PeV)	$A_{sid} \times 10^{-2}$	Phase	U.L. $\times 10^{-2}$%
5.2-5.6	2.7×10^{15}	0.26 ± 0.1	225 ± 22	0.49
5.6-6.4	6.1×10^{15}	0.29 ± 0.16	227 ± 30	0.64
≥ 6.4	3.3×10^{16}	1.2 ± 0.9	254 ± 42	3.1

Table 1. Results of harmonic analysis through the East-West method for three intervals of N_{ch}.

4. Conclusions

The KASCADE-Grande experiment took data from January 2004 to end 2012, detecting EAS generated by primary cosmic rays in the $10^{16} - 10^{18}eV$ energy range. In this contribution we have shown the main results obtained so far by the experiment:

(i) a measurement of the all-particle energy spectrum, showing that it cannot be described by a single slope power law. A hardening slightly above $10^{16}eV$ and a steepening at $\log_{10}(E/eV) = 16.92 \pm 0.10$ are detected.

(ii) The measurement of the light and heavy primary mass group energy spectra. These spectra were obtained dividing the events in two samples on the basis of the ratio between the muon and the charged particles numbers. A steepening at $\log_{10}(E/eV) = 16.92 \pm 0.04$ in the spectrum of the electron poor event sample (heavy primaries) and a hardening at $\log_{10}(E/eV) = 17.08 \pm 0.08$ in the one of the electron rich (light primaries) one were observed. The slope of the heavy mass group spectrum above the knee-like feature is similar to the one of the light mass spectrum before the the ankle-like feature.

(iii) Upper limits on the amplitude of large scale anisotropies in three N_{ch} bins.

Acknowledgments

The authors would like to thank the members of the engineering and technical staff of the KASCADE-Grande collaboration, who contributed to the success of the experiment. The KASCADE-Grande experiment is supported in Germany by the BMBF and by the Helmholtz Alliance for Astroparticle Physics - HAP funded by the Initiative and Networking Fund of the Helmholtz Association, by the MIUR and INAF of Italy, the Polish Ministry of Science and Higher Education, and the Romanian Authority for Scientific Research UEFISCDI (PNII-IDEI grants 271/2011 and 17/2011).

References

[1] K.H. Kampert and M. Unger, Astropart. Phys., **35** (2012), 660-678.
[2] R. Bonino et al., Astrophys. J. **738** (2011), 67-84.
[3] T. Antoni et al., Nucl. Instr. & Meth. A **513** (2003) 429.
[4] W.D. Apel et al., Nucl. Instr. & Meth. A **620** (2010) 202.
[5] P. Doll, et al., Nucl. Instr. & Meth. A **488** (2002) 517.
[6] W.D. Apel et al., Astropart. Phys., **34** (2011) 476485
[7] W.D. Apel et al., Astropart. Phys. **36**, 183 (2012)
[8] S. Ostapchenko, Nucl. Phys. B (Proc. Suppl.) **151** (2006) 143.
[9] L. Kuzmichev et al., Proc. 32nd Int. Cosmic Ray Conf. (Beijing) **1**, 209 (2011).
[10] M.G. Aartsen et al., Phys. Rev. D, **88**, (2013) 042004.
[11] W.D. Apel et al., J. Adv. Space Res., http//dx.doi.org/10.1016/j.asr.2013.05.008 (2013)
[12] E.-J. Ahn et al., Phys. Rev. D **80**, 094003 (2009).
[13] K. Werner. Nucl. Phys. B (Proc. Suppl.) **175** (2008) 81.
[14] S. Ostapchenko, Phys Rev D **83** (2011) 014018.
[15] T. Antoni et al., Astropart. Phys. **16**, 245 (2002).
[16] W.D. Apel et al., Phys. Rev. Lett. **107**, 171104 (2011).
[17] W.-D. Apel et al., Phys. Rev. D **87**, 081101 (2013).
[18] J. Linsley, Phys. Rev. Lett. **34** (1975) 1530.

Large-Scale Distribution of Arrival Directions of Cosmic Rays Detected at the Pierre Auger Observatory Above 10 PeV

Olivier Deligny[1], for the Pierre Auger Collaboration[2]

[1] IPN Orsay, CNRS/IN2P3 & Université Paris Sud, 15 rue Clemenceau, 91406 Orsay CEDEX, France

[2] Observatório Pierre Auger, Av. San Martín Norte 306, 5613 Malargüe, Mendoza, Argentina; `www.auger.org`; full author list: `http://www.auger.org.ar/archive/authors_2013_07.html`

E-mail: `deligny@ipno.in2p3.fr`

Abstract. Searches for large-scale anisotropies in the distribution of arrival directions of cosmic rays detected above $\simeq$ 10 PeV at the Pierre Auger Observatory are presented. Although no significant deviation from isotropy is revealed at present, some of the measurements suggest that future data will provide hints for large-scale anisotropies over a wide energy range. Those anisotropies would have amplitudes which are too small to be significantly observed within the current statistics. Assuming that the cosmic ray anisotropy is dominated by dipole and quadrupole moments in the EeV-energy range, some consequences of the present upper limits on their amplitudes are presented.

1. Introduction

The large-scale distribution of arrival directions of cosmic rays is an important observable in attempts to understand their origin. This is because this observable is closely connected to both their source distribution and their propagation. Due to scattering in magnetic fields, the anisotropy imprinted in the arrival directions is mainly expected at large scales up to the highest energies. For energies above a few PeV, the decrease of the intensity with energy makes it challenging to collect the necessary statistics required to reveal amplitudes down to $10^{-3} - 10^{-2}$.

The surface detector arrays operating at the Pierre Auger Observatory provide high-quality data to search for large-scale anisotropies from $\simeq$10 PeV up to the highest energies. The different analyses performed by the Pierre Auger Collaboration on data taken from 2004 until the end of 2012 are summarised in this paper. Comprehensive reports describing all the details of the analyses can be found in [1, 2, 3, 4, 5, 6, 7]. An emphasis is given here to hints that may be indicative of anisotropies whose amplitudes are still too small to stand out significantly above the background noise which is intrinsic to the finite statistics accumulated so far, and to the implications on the origin of EeV-cosmic rays that can be inferred from the upper limits obtained in this energy range.

The 1500 m surface detector array of the Pierre Auger Observatory is composed of 1600 water-Cherenkov detectors arranged in a triangular grid with spacing of 1500 m, covering a total area of 3000 km^2. This array is fully efficient at energies above 3 EeV. 49 additional detectors with 750 m spacing have been nested within the 1500 m array to cover an area of 25 km^2 with full efficiency above 300 PeV. This 750 m surface detector array allows complementary anisotropy searches at much lower energies, down to $\simeq$ 10 PeV.

2. Harmonic Analyses in Right Ascension Above 10 PeV

2.1. Event Rate

To search for large-scale anisotropies, it is mandatory to account for the modulation of the event rate imprinted by variations of experimental origin. High-quality events used here are selected on the condition that the elemental cell of the event (that is, all six neighbors of the water-Cherenkov detector with the highest signal) was active at the time the event was recorded [8]. Based on this fiducial cut, the total geometric exposure to the end of 2012 was 31645 km^2 sr yr for the 1500 m surface detector array and 79 km^2 sr yr for the 750 m one. Since the number of elemental cells $n_{\mathrm{cell}}(t)$ is accurately monitored every second, the expected number of events for a uniform flux can be inferred at the corresponding accuracy as a function of time. Denoting by α^0 the local sidereal time and by T_{sid} the sidereal period, the total number of elemental cells $N_{\mathrm{cell}}(\alpha^0)$ and its associated relative variations $\Delta N_{\mathrm{cell}}(\alpha^0)$ can then be obtained as:

$$N_{\mathrm{cell}}(\alpha^0) = \sum_j n_{\mathrm{cell}}(\alpha^0 + jT_{\mathrm{sid}}), \qquad \Delta N_{\mathrm{cell}}(\alpha^0) = \frac{N_{\mathrm{cell}}(\alpha^0)}{\langle N_{\mathrm{cell}}\rangle_{\alpha^0}}, \tag{1}$$

with $\langle N_{\mathrm{cell}}\rangle_{\alpha^0} = 1/T_{\mathrm{sid}} \int_0^{T_{\mathrm{sid}}} \mathrm{d}\alpha^0 N_{\mathrm{cell}}(\alpha^0)$. This latter quantity is the relevant input to account for the changes in time of the observatory when estimating the anisotropy parameters.

2.2. Event Rate and Weather Effects

The energy estimator of the showers collected at the surface detector array of the Pierre Auger Observatory is provided by the signal size measured at 1000 m from the shower core, $S(1000)$. Since the development of extensive air showers depends on the atmospheric pressure P and air density ρ, the corresponding $S(1000)$ for any fixed energy is sensitive to variations in pressure and air density. Systematic variations with time of $S(1000)$ induce variations of the event rate that may distort the real dependence of the cosmic ray intensity with right ascension. To cope with this experimental effect, the observed shower size $S(1000)$, measured at the actual density ρ and pressure P, is converted to the one that would have been measured at reference values ρ_0 and P_0 [9]. Applying these corrections to the energy assignments of showers allows the cancellation of spurious variations of the event rate in right ascension, whose typical amplitudes amount to a few per thousand when considering data sets collected over full years [1].

2.3. Different Analyses

To account for the non-uniform exposure in the sky, the inverse of $\Delta N_{\mathrm{cell}}(\alpha^0)$ can be used to weight the contribution of each event entering in the calculation of the Fourier first-harmonic coefficients [10]:

$$a = \sum_{i=1}^{N} \frac{\cos\alpha_i}{\Delta N_{\mathrm{cell}}(\alpha_i^0)}, \qquad b = \sum_{i=1}^{N} \frac{\sin\alpha_i}{\Delta N_{\mathrm{cell}}(\alpha_i^0)}. \tag{2}$$

The sum runs over the number N of events in the energy range considered, and the normalisation reads as $\mathcal{N} = \sum_{i=1}^{N} 1/\Delta N_{\mathrm{cell}}(\alpha_i^0)$. The amplitude and phase are then calculated as $r = \sqrt{a^2+b^2}$ and $\phi = \arctan(b/a)$. For isotropy, and for weight values close to 1, these random variables follow a Rayleigh and a uniform distributions respectively [11].

This procedure has been shown to provide accurate estimates for data collected at the 1500 m surface detector array above 1 EeV [1]. Below 1 EeV, in addition to the energy estimates, weather effects can also affect the detection efficiency as a function of time and thus imprint fake anisotropies in the arrival directions. To circumvent this difficulty, the estimation of the anisotropy parameters is then carried out using the East-West method [12], which is based on the fact that the instantaneous exposure for Eastward and Westward events is the same. Hence, the difference between the event counting rate measured from the East sector and the West sector can provide unbiased estimates of anisotropy parameters without applying any correction, although at the price of reducing the sensitivity to the first harmonic modulation. Note also that, in the case of the 750 m surface detector array, results are presented only in terms of the East-West method.

2.4. Measurements of First Harmonic Amplitudes

The Rayleigh amplitude r measured by any observatory can be used to reveal anisotropies projected on the Earth's equatorial plane. In the case of an underlying pure dipole, the relationship between r and the projection of the dipole on the Earth's equatorial plane, $d_\perp$, depends on the latitude of the observatory and on the range of zenith angles considered: $d_\perp \simeq r/\langle\cos\delta\rangle$. $d_\perp$ is thus the physical quantity of interest to compare the results of different experiments and the pure dipole predictions.

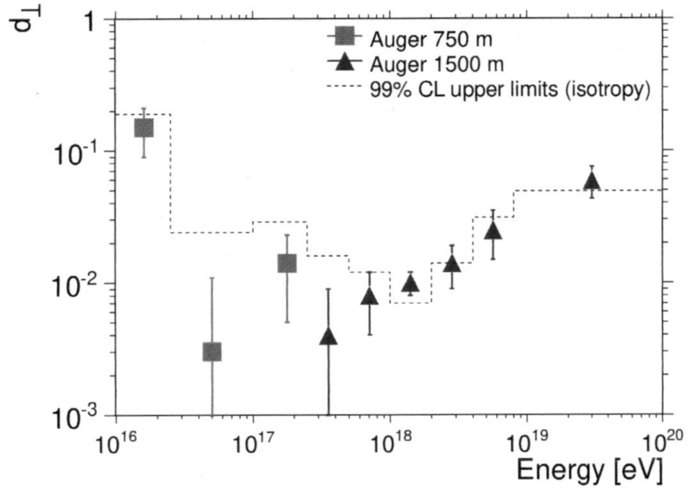

Figure 1. First harmonic amplitude of the equatorial dipole anisotropy as a function of energy. The dashed lines are the 99% confidence level upper values of the amplitude that could result from fluctuations of an isotropic distribution. From [4].

The amplitudes $d_\perp$ obtained are shown in figure 1. The dashed line stands for the upper values of the amplitude which may arise from fluctuations in an isotropic distribution at 99% confidence level. Note that in the energy ranges 1-2 and 2-4 EeV the measured amplitudes have a probability to arise by chance from an isotropic distribution of about 0.03% and 0.9%, while above 8 EeV the measured amplitude has chance probability of only 0.1%. Since several energy bins were searched, these numbers do not represent absolute probabilities. They constitute however interesting hints for large-scale anisotropies that will be important to further scrutinise with increased statistics.

2.5. Notes on the Measurements of First Harmonic Phases

It has been noted in the past that, for an underlying anisotropy, the detection of the genuine phase is expected to occur earlier than the detection of a significant amplitude [13]. For an anisotropy evolving smoothly with the energy, testing the consistency of independent phase measurements, such as a constancy or a smooth evolution in adjacent energy bins, may thus be a powerful tool for detecting anisotropy.

A likelihood ratio test can be designed to quantify whether or not a parent random distribution of arrival directions better reproduces the phase measurements in different energy intervals than an alternative dipolar parent distribution [1]. The likelihood ratio λ is built from the p.d.f. of each independent measurement (let's suppose here N_b independent bins ordered in energy, with $N = 30,000$ events in each bin) under the hypothesis of isotropy ($p_\Phi^{\text{iso}}(\phi)$, uniform distribution) and under the hypothesis of a dipolar distribution ($p_\Phi^{\text{dipole}}(\phi)$ calculated for the first time in [11]):

$$\lambda = \frac{\prod_{i=1}^{i=N_b} p_\Phi^{\text{iso}}(\phi_i)}{\prod_{i=1}^{i=N_b} p_\Phi^{\text{dipole}}(\phi_i)}. \tag{3}$$

The expected amplitude values s entering into each $p_\Phi^{\text{dipole}}(\phi_i)$ can be estimated by the expected mean noise given the available statistics. The statistic of the variable $-2\ln(\lambda)$ is then built under the hypothesis of isotropy by means of a large number of Monte-Carlo isotropic samples. The probability that the hypothesis of isotropy better reproduces the measurements compared to the alternative hypothesis is then calculated by integrating the normalised distribution of $-2\ln(\lambda)$ above the value found in the data set.

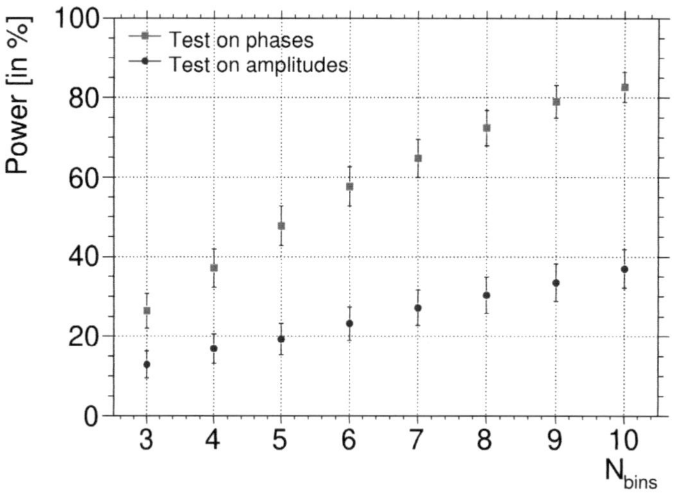

Figure 2. Power of the tests on amplitudes (in blue) and on phases (in red) as a function of the number of bins N_b entering in each test, in the case of a genuine signal $s = 1\%$ and with $N = 30,000$ events in each bin.

The power of this test is illustrated in figure 2, where its efficiency (shown as the red squares) to detect a genuine anisotropy for a threshold value of 1% is shown as a function of the number of bins N_b. Compared to the efficiency of the "$2K$" test on amplitudes (see [13] for details about the "$2K$" test), it is apparent that the test on the consistency of the phase measurements leads to a better power by a factor greater than 2.

Driven by this property, it was pointed out in [1] that the phase measurements in adjacent energy intervals above $\simeq$ 100 PeV was suggestive of a smooth transition between $\simeq 270°$ below 1 EeV and another phase ($\simeq 100°$) above 5 EeV. This behaviour motivated the Pierre Auger Collaboration to design a prescription with the intention of establishing at 99% confidence level whether this consistency in phases in adjacent energy intervals is real or not. The test makes use of all high-quality events above 10 PeV collected at the 750 m surface detector array, and all

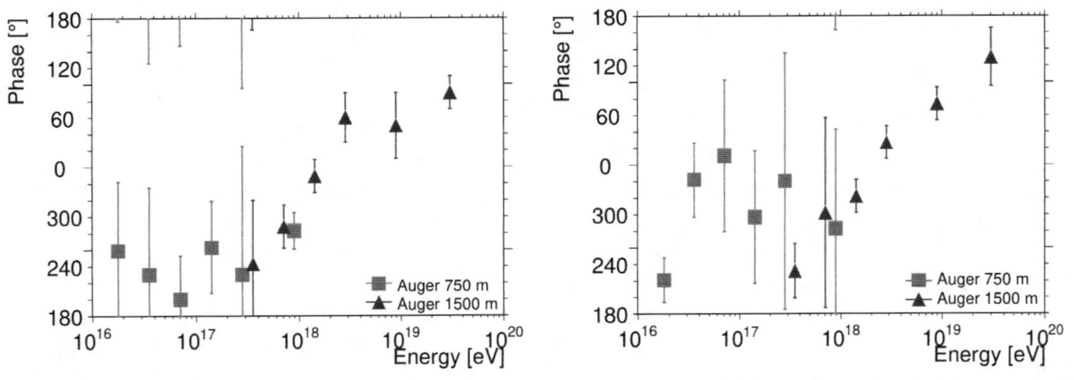

Figure 3. Phase of the first harmonic in right ascension as a function of energy. Left: Data collected from January 1 2004 to December 31 2010 for the 1500 m array, and from September 12 2007 to April 11 2011 for the 750 m one. Right: Data from June 25 2011 (beginning of the prescription) up to December 31 2012. From [4].

high-quality events above 250 PeV collected at the 1500 m surface detector array. The mid-term status of the prescription, reported in [4], is illustrated in figure 3. The right panel is derived with data collected since June 25 2011 (beginning of the prescription) up to December 31 2012 only. At this stage, the values as derived from the analysis applied to the 750 m array are still affected by large uncertainties. On the other hand, the overall behavior of the points as derived from the analysis applied to the 1500 m array shows good agreement with the previous observations. The final result of the prescription is expected for 2015, once the required exposure in the prescription is reached.

3. Spherical Harmonic Analyses Above 1 EeV

3.1. Event Rate and Geomagnetic Field

To characterise the arrival directions in both right ascension and declination angles, it is essential to keep under control the observed rate in terms of local angles. The geomagnetic field turns out to influence the shower developments and shower size estimations at a fixed energy due to the broadening of the spatial distribution of particles in the direction of the Lorentz force. As the strength of the geomagnetic field component perpendicular to any arrival direction depends on both the zenith and azimuthal angles, the small changes of the density of particles at ground break the circular symmetry of the lateral spread of the particles and thus induce a dependence of the shower size $S(1000)$ at a fixed energy in terms of the azimuthal angle. Due to the steepness of the energy spectrum, such an azimuthal dependence translates into azimuthal modulations of the observed cosmic ray event rate at a given $S(1000)$. To eliminate these effects, the observed shower size $S(1000)$ is thus converted to the one that would have been observed in the absence of geomagnetic field [14].

3.2. Directional Exposure

The *directional exposure* ω of the Observatory, providing the effective time-integrated collecting area for a flux from each direction of the sky in unitsof km^2 yr, is the key-ingredient to be determined accurately in searching for anisotropies. The directional aperture per cell can be known in a pure geometric way as:

$$a_{\text{cell}}^{(i)}(\theta,\varphi) = 1.95\ [1+\zeta^{(i)}\tan\theta\cos(\varphi-\varphi_0^{(i)})]\ \cos\theta\ \text{km}^2, \tag{4}$$

Journal of Physics: Conference Series **531** (2014) 012002 doi:10.1088/1742-6596/531/1/012002

where $\zeta^{(i)}$ and $\varphi_0^{(i)}$ are the zenith and azimuth angles of the normal vector to each elemental cell. Given that the number of high-quality selected events is proportional to the number $n_{\rm cell}(t)$ of available cells at any time, the directional exposure is controlled by this number $n_{\rm cell}(t)$ [8]. In addition, for energies below 3 EeV, it is also controlled by the detection efficiency ϵ for triggering. This efficiency depends on the energy E, the zenith angle θ, and the azimuth angle φ.

To determine the detection efficiency function, an empirical approach based on the quasi-invariance of the zenith distribution to large-scale anisotropies for zenith angles less than $\simeq 60°$ and for any Observatory whose latitude is far from the poles of the Earth can be adopted [3]. For full efficiency, the distribution $dN/d\sin^2\theta$ is expected to be uniform for geometry reasons. Consequently, below full efficiency, deviations from a uniform behaviour in the $dN/d\sin^2\theta$ distribution provides an empirical measurement of the zenith dependence of the detection efficiency. Since this is applied to all events detected without distinction based on the primary mass of cosmic rays, this technique does not provide the mass dependence of the detection efficiency. For that reason, the anisotropy searches reported in [3, 5] and presented below pertain to the whole population of cosmic rays, whether this population consists of a single primary mass or a mixture of several elements.

Because the observed azimuthal distribution $dN/d\varphi$ is, in contrast, sensitive to large-scale anisotropies [3], the same empirical approach cannot be used to determine the variations with φ around the mean value of ϵ previously derived for each energy and each zenith angle. The effects inducing these variations have thus to be modelled accurately.

The first effect is due to the influence of the geomagnetic field. Indeed, if the procedure consisting in correcting the energies for the geomagnetic effects guarantees an unbiased event rate for full efficiency, it does not account for the fact that under any incident angles (θ, φ) below full efficiency, the observed showers triggered the surface detector array with a probability associated to their *uncorrected* sizes. Above 1 EeV, this effect is the main source of azimuthal dependence of the detection efficiency. It can be accounted for in a straightforward way by retro-propagating the angular dependence of the geomagnetic corrections into the energy given in the argument of the mean ϵ function determined previously.

The second relevant effect is induced by the tilt of the array. This is because the effective separation between detectors for a given zenith angle depends actually on the azimuth. Since the surface detector array seen by showers coming from the uphill direction is denser than that for those coming from the downhill one, the detection efficiency is higher in the uphill direction. The corresponding change in detection efficiency $\Delta\epsilon_{\rm tilt}(\theta, \varphi, E)$ can be parameterised in a quite generic way [3].

Overall, the final expression used to calculate the directional exposure is the following:

$$\omega(\delta, E) = \sum_{i=1}^{n_{\rm cell}} x^{(i)} \int_0^{24h} \mathrm{d}\alpha' \, a_{\rm cell}^{(i)}(\theta, \varphi) \; [\epsilon(\theta, \varphi, E) + \Delta\epsilon_{\rm tilt}(\theta, \varphi, E)] \,, \tag{5}$$

where $x^{(i)}$ is the total operational time of the cell (i) considered as independent of the local sidereal time [1]. The integration variable α' stands for the hour angle. The transformation from local to celestial angles accounts for the slightly different latitudes of the cells to account for the spatial extension of the surface detector array. For a wide range of declinations between $\simeq -89°$ and $\simeq 20°$, the directional exposure is $\simeq 2,990$ km^2 yr at 1 EeV and $\simeq 4,180$ km^2 yr above 3 EeV.

[1] Although the operational time of each cell has not been constant within each sidereal day due to temporary instabilities and/or failures, the resulting modulation is overall negligible once the total operational time obtained by summing over a large number of sidereal days is considered.

3.3. Estimates of Spherical Harmonic Coefficients

Any angular distribution over the sphere $\Phi(\mathbf{n})$ can be decomposed in terms of a multipolar expansion:

$$\Phi(\mathbf{n}) = \sum_{\ell \geq 0} \sum_{m=-\ell}^{\ell} a_{\ell m} Y_{\ell m}(\mathbf{n}), \tag{6}$$

where $\mathbf{n}$ denotes a unit vector taken in equatorial coordinates. The customary recipe to extract each multipolar coefficient makes use of the completeness relation of spherical harmonics:

$$a_{\ell m} = \int_{4\pi} \mathrm{d}\Omega\ \Phi(\mathbf{n}) Y_{\ell m}(\mathbf{n}), \tag{7}$$

where the integration is over the entire sphere of directions $\mathbf{n}$. Any anisotropy fingerprint is encoded in the $a_{\ell m}$ spherical harmonic coefficients. Variations on an angular scale of Θ radians contribute amplitude in the $\ell \simeq 1/\Theta$ modes. However, due the non-uniform and incomplete coverage of the sky at the Pierre Auger Observatory, the estimated coefficients $\overline{a}_{\ell m}$ are determined in a two-step procedure. First, from any event set with arrival directions $\{\mathbf{n_1}, ..., \mathbf{n_N}\}$ recorded at local sidereal times $\{\alpha_1^0, ..., \alpha_N^0\}$, the multipolar coefficients of the angular distribution coupled to the exposure function are estimated through:

$$\overline{b}_{\ell m} = \sum_{k=1}^{N} \frac{Y_{\ell m}(\mathbf{n}_k)}{\Delta N_{\mathrm{cell}}(\alpha_k^0)}. \tag{8}$$

The weights $1/\Delta N_{\mathrm{cell}}(\alpha_k^0)$ are introduced to correct for the slightly non-uniform directional exposure in right ascension. Then, assuming that the multipolar expansion of the angular distribution $\Phi(\mathbf{n})$ is *bounded* to ℓ_{max}, the first $b_{\ell m}$ coefficients with $\ell \leq \ell_{\mathrm{max}}$ are related to the non-vanishing $a_{\ell m}$ through:

$$\overline{b}_{\ell m} = \sum_{\ell'=0}^{\ell_{\mathrm{max}}} \sum_{m'=-\ell'}^{\ell'} [K]_{\ell m}^{\ell' m'} \overline{a}_{\ell' m'}, \tag{9}$$

where the matrix K is entirely determined by the directional exposure:

$$[K]_{\ell m}^{\ell' m'} = \int_{\Delta\Omega} \mathrm{d}\Omega\ \omega(\mathbf{n})\ Y_{\ell m}(\mathbf{n})\ Y_{\ell' m'}(\mathbf{n}). \tag{10}$$

Inverting Eqn. 9 allows a recovering of the underlying $\overline{a}_{\ell m}$, with a resolution proportional to $([K^{-1}]_{\ell m}^{\ell m}\ \overline{a}_{00})^{0.5}$ [15]. As a consequence of the incomplete coverage of the sky, this resolution deteriorates by a factor larger than 2 each time ℓ_{max} is incremented by 1. Given the current present statistics, this prevents the recovery of each coefficient with good accuracy as soon as $\ell_{\mathrm{max}} \geq 3$.

3.4. Dipole and Quadrupole Moments

For the reason explained just above, the angular distribution $\Phi(\mathbf{n})$ is assumed to be modulated by a dipole and a quadrupole only. In this case, $\Phi(\mathbf{n})$ can be written in a convenient geometric way as:

$$\Phi(\mathbf{n}) = \frac{\Phi_0}{4\pi} \left(1 + r\ \mathbf{d} \cdot \mathbf{n} + \lambda_+(\mathbf{q}_+ \cdot \mathbf{n})^2 + \lambda_0(\mathbf{q}_0 \cdot \mathbf{n})^2 + \lambda_-(\mathbf{q}_- \cdot \mathbf{n})^2\right). \tag{11}$$

The dipole pattern is here fully characterised by a vector with declination δ_d, right ascension α_d, and an amplitude r corresponding to the maximal anisotropy contrast: $r = (\Phi_{\mathrm{max}} - \Phi_{\mathrm{min}})/(\Phi_{\mathrm{max}} + \Phi_{\mathrm{min}})$. On the other hand, the quadrupolar pattern is fully characterised by the eigenvalues and the corresponding eigenvectors of a tensor. It is convenient to define the

Journal of Physics: Conference Series **531** (2014) 012002 doi:10.1088/1742-6596/531/1/012002

quadrupole amplitude $\beta \equiv (\lambda_+ - \lambda_-)/(2 + \lambda_+ + \lambda_-)$, which provides a measure of the maximal quadrupolar contrast in the absence of a dipole. The quadrupole is then described in terms of two amplitudes (β, λ_+) and three angles: (δ_+, α_+) which define the orientation of $\mathbf{q}_+$ and (α_-) which defines the direction of $\mathbf{q}_-$ in the orthogonal plane to $\mathbf{q}_+$. The third eigenvector $\mathbf{q}_0$ is orthogonal to $\mathbf{q}_+$ and $\mathbf{q}_-$, and has an amplitude determined from the traceless constraint $\lambda_+ + \lambda_- + \lambda_0 = 0$.

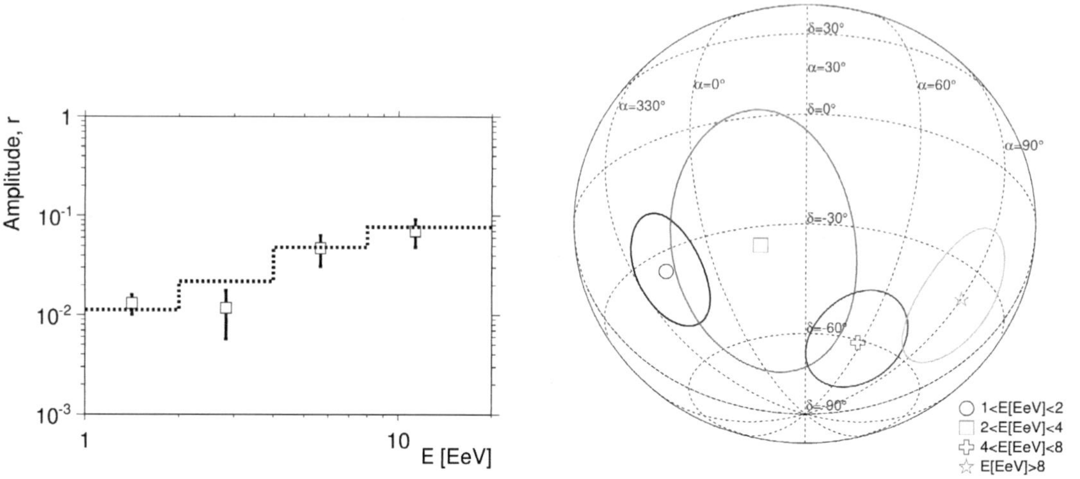

Figure 4. Left: Reconstructed amplitude of the dipole as a function of the energy. The dotted line stands for the 99% confidence level upper bounds on the amplitudes that would result from fluctuations of an isotropic distribution. Right: Reconstructed declination and right ascension of the dipole with corresponding uncertainties, as a function of the energy, in orthographic projection. From [5].

For a pure dipole ($\lambda_{\pm,0} = 0$), the reconstructed amplitudes are shown in figure 4 as a function of the energy. The 99% confidence level upper bounds on the amplitudes that would result from fluctuations of an isotropic distribution are indicated by the dotted line. Similarly to the results presented in last section, interesting hints for large-scale anisotropies are observed in the first energy bin, hints that will be important to further scrutinise with independent data. The corresponding reconstructed directions in orthographic projection with the associated uncertainties are also shown as a function of the energy. The same behaviour is observed for the phases as in the previous analysis. Considering also the possibility for a non-zero quadrupole, the dipole amplitude in the first energy bin is not any longer above the corresponding 99% confidence level upper bound due to the lost of sensitivity when increasing the upper bound of the multipolar expansion.

The upper limits that can be derived on the dipole and quadrupole amplitudes are presented and discussed in the next section.

4. Constraints on the Origin of EeV-Cosmic Rays

From these analyses, the upper limits obtained at 99% confidence level on dipole and quadrupole amplitudes are shown in figure 5. These upper limits are stringent enough to challenge some Galactic scenarios. One of them, in which sources of EeV-cosmic rays are stationary, densely and uniformly distributed in the Galactic disk, and emit particles in all directions, is illustrated in the following.

Both the strength and the structure of the magnetic field in the Galaxy, known only approximately, are of central importance for the propagation of EeV-cosmic rays. While the turbulent component dominates in strength by a factor of a few, the regular component, of the order of few microgauss in the disk, is expected to imprint dominant drift motions as soon as the Larmor radius of cosmic rays is larger than the maximal scale of the turbulences - thought to be in the range 10-100 pc. The parameterisation adopted here is the one obtained recently in [16]. Note however that estimations based on a much improved model reported in [17] are currently being carried out [18]. Preliminary results are overall in good agreement in terms of the expected anisotropies with the ones reported in [2, 3] and presented here.

To describe the propagation of EeV-cosmic rays in such a magnetic field, the direct integration of trajectories is the most appropriate tool. To obtain the anisotropy of cosmic rays emitted from sources uniformly distributed in a cylinder with a radius of 20 kpc from the galactic centre and with a height of $\pm$ 100 pc, the widely-used method first proposed in [19] and adopted in [2, 3] consists in back-tracking anti-particles with random directions from the Earth to outside the Galaxy. Each test particle *probes* the total luminosity along the path of propagation from each direction as seen from the Earth. For *stationary sources emitting cosmic rays in all directions*, the expected flux in the initial sampled direction is proportional to the time spent by each test particle in the source region.

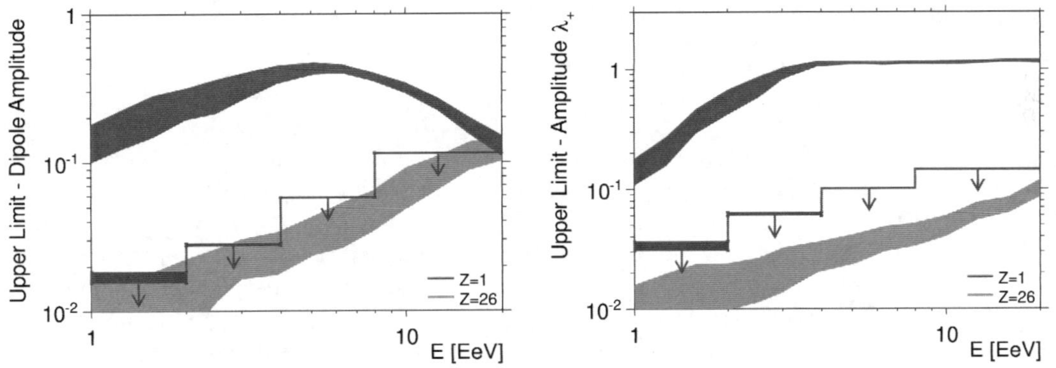

Figure 5. 99% *C.L.* upper limits on dipole and quadrupole amplitudes as a function of the energy. Some generic anisotropy expectations from stationary Galactic sources distributed in the disk are also shown, for protons and iron-nuclei. The fluctuations of the amplitudes due to the stochastic nature of the turbulent component of the magnetic field are sampled from different simulation data sets and are shown by the bands (see text).

The dipole and quadrupole amplitudes obtained for several energy values covering the range $1 \leq E/\mathrm{EeV} \leq 20$ are shown in figure 5. Simulated event sets with 5 10^5 test particles were generated to shrink the statistical uncertainties on amplitudes at the 0.5% level - which is necessary to probe unambiguously amplitudes down to the percent level. The shaded regions stand for the RMS of the amplitudes sampled from 20 independent sets of 5 10^5 test particles, where the stochastic component of the magnetic field (*i.e.* the turbulence) is frozen differently in each set. Clearly, the resulting amplitudes for protons stand well above the allowed limits. Consequently, unless the strength of the magnetic field is much higher than in the picture used here, the upper limits derived in this analysis exclude the light component of cosmic rays coming from galactic stationary sources densely distributed in the galactic disk and emitting in all directions. To respect the dipole limits below the ankle energy, the fraction of protons should not exceed $\simeq$ 10% of the cosmic-ray composition unless these protons do not originate

Journal of Physics: Conference Series **531** (2014) 012002 doi:10.1088/1742-6596/531/1/012002

from sources that can be roughly characterised by the picture used here. This is particularly interesting in the view of the indications for the presence of a light component around 1 EeV from shower depth of maximum measurements [20, 21, 22, 23], though firm interpretations of these measurements in terms of the atomic mass still suffer from some ambiguity due to the uncertain hadronic interaction models used to describe the shower developments.

5. Large-Scale Structure with Full-Sky Coverage Above 10 EeV

As stressed in section 3, the only way to unambiguously measure multipole moments to any order requires full-sky coverage. At present, full-sky coverage can only be achieved through the meta-analysis of data recorded at observatories located in both hemispheres of the Earth. The Telescope Array is the largest experiment ever built in the Northern hemisphere to study ultra-high energy cosmic rays. Given the respective latitudes and zenith ranges covered by both experiments, full-sky coverage can be achieved by combining both Pierre Auger Observatory and Telescope Array data sets. To facilitate the first joint analysis of this kind, the energy threshold, 10 EeV, is chosen to guarantee that both observatories operate with full-detection efficiency [6, 7]. In this way, the respective exposure functions follow purely geometric expectations.

The main challenge in combining the data sets is to account adequately for the relative exposures of both experiments. In principle, the combined directional exposure of the two experiments should be simply the sum of the individual ones. However, individual exposures have here to be re-weighted by some empirical factor b due to the unavoidable uncertainty in the relative exposures of the experiments:

$$\omega(\mathbf{n}; b) = \omega_{\mathrm{TA}}(\mathbf{n}) + b\omega_{\mathrm{Auger}}(\mathbf{n}). \qquad (12)$$

Written in this way, b is a dimensionless parameter of order unity. In practice, only an estimation $\overline{b}$ of the factor b can be obtained, so that only an estimation of the directional exposure $\overline{\omega}(\mathbf{n}) \equiv \omega(\mathbf{n}; \overline{b})$ can be achieved through equation 12. The parameter b can thus be viewed as a fudge factor which absorbs any kind of systematic uncertainties in the relative exposures, whatever the sources of these uncertainties - resulting from different energy scales for instance. This empirical factor is arbitrary chosen in the following to re-weight the directional exposure of the Pierre Auger Observatory relatively to the one of the Telescope Array.

A band of declinations around the equatorial plane is exposed to the fields of view of both experiments, namely for declinations between $-15°$ and $25°$. This overlapping region can be used for designing an empirical procedure to get a relevant estimate of the parameter b [6]. Considering as a first approximation the flux $\Phi(\mathbf{n})$ as isotropic, the overlapping region $\Delta\Omega$ can be utilised to derive a first estimate $\overline{b}^{(0)}$ of the b factor by *forcing the intensities of both experiments to be identical in this particular region.* This can be easily achieved in practice, by taking the ratio of the ΔN_{TA} and $\Delta N_{\mathrm{Auger}}$ events observed in the overlapping region $\Delta\Omega$ weighted by the ratio of nominal exposures:

$$\overline{b}^{(0)} = \frac{\Delta N_{\mathrm{Auger}}}{\Delta N_{\mathrm{TA}}} \frac{\int_{\Delta\Omega} d\Omega\ \omega_{\mathrm{TA}}(\mathbf{n})}{\int_{\Delta\Omega} d\Omega\ \omega_{\mathrm{Auger}}(\mathbf{n})}. \qquad (13)$$

Then, inserting $\overline{b}^{(0)}$ into $\overline{\omega}$, 'zero-order' $\overline{a}_{\ell m}^{(0)}$ coefficients can be obtained. This set of coefficients is only a rough estimation, due to the limiting assumption on the flux (isotropy).

On the other hand, the expected number of events in the common band for each observatory, $\Delta n_{\mathrm{TA}}^{\mathrm{exp}}$ and $\Delta n_{\mathrm{Auger}}^{\mathrm{exp}}$, can be expressed from the underlying flux $\Phi(\mathbf{n})$ and the true value of b as:

$$\Delta n_{\mathrm{TA}}^{\mathrm{exp}} = \int_{\Delta\Omega} d\Omega\ \Phi(\mathbf{n})\omega_{\mathrm{TA}}(\mathbf{n})$$

$$\Delta n_{\rm Auger}^{\rm exp} \quad = \quad b\int_{\Delta\Omega} d\Omega\ \Phi(\mathbf{n})\omega_{\rm Auger}(\mathbf{n}). \tag{14}$$

From equations 14, and from the set of $\overline{a}_{\ell m}^{(0)}$ coefficients, an iterative procedure estimating at the same time b and the set of $a_{\ell m}$ coefficients can be considered in the following way:

$$\overline{b}^{(k+1)} = \frac{\Delta N_{\rm Auger}}{\Delta N_{\rm TA}} \frac{\int_{\Delta\Omega} d\Omega\ \overline{\Phi}^{(k)}(\mathbf{n})\omega_{\rm TA}(\mathbf{n})}{\int_{\Delta\Omega} d\Omega\ \overline{\Phi}^{(k)}(\mathbf{n})\omega_{\rm Auger}(\mathbf{n})}, \tag{15}$$

where $\Delta N_{\rm TA}$ and $\Delta N_{\rm Auger}$ as derived in the first step are used to estimate $\Delta n_{\rm TA}^{\rm exp}$ and $\Delta n_{\rm Auger}^{\rm exp}$ respectively, and $\overline{\Phi}^{(k)}$ is the flux estimated with the set of $\overline{a}_{\ell m}^{(k)}$ coefficients.

This procedure makes it possible to use the powerful multipolar analysis method for characterising the sky map of ultra-high energy cosmic rays. It pertains to any full-sky coverage achieved by combining data sets from different observatories, and opens a rich field of anisotropy studies. This technique will be applied to data sets from the Telescope Array and the Pierre Auger Observatory and will be reported in a near future [7].

6. Outlook

A thorough search for large-scale anisotropies in the distribution of arrival directions of cosmic rays detected above $\simeq$ 10 PeV at the Pierre Auger Observatory has been presented. With respect to the traditional search in right ascension only, spherical harmonic analyses has been performed above 1 EeV to characterise the distribution of arrival directions in terms of both the right ascension and the declination. No significant deviation from isotropy is revealed within the systematic uncertainties, although there are few hints that may be indicative of a genuine signal whose amplitude is at the level of the current statistical noise. The current sensitivity allows us to challenge an origin of cosmic rays from stationary galactic sources densely distributed in the galactic disk and emitting predominantly light particles in all directions.

Future work will profit from the increased statistics. Also, future joint studies between the Telescope Array and the Pierre Auger Collaborations will allow a determination of the full set of spherical harmonic coefficients at ultra-high energies. This will provide further constraints helping to understand the origin of cosmic rays above $\simeq$ 10 PeV.

Acknowledgements

The author thanks the organisers of the workshop for the opportunity to present these results and the members of the Pierre Auger Collaboration for all the work described in this contribution.

References

[1] The Pierre Auger Collaboration 2011, Astropart. Phys., 34 628
[2] The Pierre Auger Collaboration 2012, ApJL, 762, L13
[3] The Pierre Auger Collaboration 2012, ApJS, 203 34
[4] I. Sidelnik for the Pierre Auger Collaboration, Proceedings of the 33rd ICRC, 2013, Rio de Janeiro, Brazil, `arXiv:1307.5059`
[5] R. Menezes for the Pierre Auger Collaboration, Proceedings of the 33rd ICRC, 2013, Rio de Janeiro, Brazil, `arXiv:1307.5059`
[6] O. Deligny for the Telescope Array and Pierre Auger Collaborations, Proceedings of the 33rd ICRC, 2013, Rio de Janeiro, Brazil, `arXiv:1310.0647`
[7] The Telescope Array and Pierre Auger Collaborations, in preparation
[8] The Pierre Auger Collaboration 2010, Nucl. Instr. and Meth. A 613 29
[9] The Pierre Auger Collaboration 2011, Astropart. Phys., 32 89
[10] S. Mollerach, E. Roulet 2005, JCAP 0508 004

Journal of Physics: Conference Series **531** (2014) 012002 doi:10.1088/1742-6596/531/1/012002

[11] J. Linsley 1975, Phys. Rev. Lett., 34 1530
[12] R. Bonino *et al.* 2011, ApJ 738 67
[13] D. Edge *et al.* 1978, J. Phys. G 133
[14] The Pierre Auger Collaboration 2011, JCAP 11 022
[15] P. Billoir, O. Deligny 2008, JCAP 02 009
[16] M. S. Pshirkov *et al.* 2011, ApJ 738 192
[17] R. Jansson, G. R. Farrar 2012, ApJL 761 L11
[18] N. Awal, G. Farrar 2014, in preparation
[19] K. O. Thielheim, W. Langhoff 1968, J. Phys. A 694
[20] The Pierre Auger Collaboration 2010, Phys. Rev. Lett. 104 091101
[21] R. U. Abbasi *et al.* (the HiRes Collaboration) 2010, Phys. Rev. Lett. 104 161101
[22] C. C. H. Jui *et al.* (the Telescope Array Collaboration) 2011, Proceedings of the APS meeting, arXiv:1110.0133
[23] E. G. Berezhko *et al.* 2012, Astropart. Phys. 36 31

Galactic magnetic deflections of UHECRs including realistic random fields

Azadeh Keivani[1], **Glennys Farrar**[2], **Michael Sutherland**[1,3,4] **and James Matthews**[1]

[1]Department of Physics & Astronomy, Louisiana State University, Baton Rouge, LA
[2]Center for Cosmology and Particle Physics, New York University, New York, NY
[3]Department of Physics, The Ohio State University, Columbus, OH
[4]Center for Cosmology and AstroParticle Physics, The Ohio State University, Columbus, OH

E-mail: akeivani@phys.lsu.edu

Abstract. We present the results of a study that simulates trajectories of ultra-high energy cosmic rays from Centaurus A to Earth. The arrival directions are characterized to assess whether Cen A can be identified as a source of cosmic rays based on observing departures from isotropy. We analyze separately particles originating from the central engine as well as from the north and south radio lobes of the Cen A complex. Simulations are performed for particle rigidities from $E/Z = 2$ EV to 100 EV, thus covering the possibility of primary particles as heavy as Fe nuclei with energies exceeding 50 EeV. The Galactic magnetic field is modeled using the recent work of Jansson and Farrar (JF12) which fitted its parameters to match extragalactic Faraday rotation measures and WMAP7 synchrotron emission maps. We also discuss the effects of an additional turbulent magnetic field on the cosmic ray deflections.

1. Introduction

One of the biggest mysteries in particle astrophysics is the origin of ultra-high energy cosmic rays (UHECRs). There are few source types that appear capable of accelerating the particles to energies as high as has been observed ($> 10^{20}$ eV). Moreover, UHECRs with energies above 5×10^{19} eV (50 EeV) lose energy through interactions with cosmic microwave background photons so that any that are observed must have originated within about 200 Mpc [1, 2]. Centaurus A (Cen A) is the closest active galactic nucleus (AGN) to Earth and is one of these source candidates. The Pierre Auger Observatory has reported a modest excess of cosmic ray arrival directions from within about 20° of the celestial location of Cen A in their UHECR data [3]. This greatly increases the interest to study the propagation of UHECRs through the Galactic magnetic fields (GMF) from Cen A.

UHECRs are mostly or entirely charged particles [4] so they will be deflected in the magnetic fields that they traverse on their way to Earth. In this study, we simulate trajectories of cosmic rays from Cen A to Earth to examine the deflections that occur due to the GMF. We construct the GMF according to the new model of Jansson and Farrar [5, 6] (JF12). The parameters of the JF12 model were tuned to give an excellent global fit to observed rotation measures and polarized synchrotron data. An overview of the model is given in Section 2.

This study is an extension to the work of Farrar *et al.* [7], where they simulated UHECRs from center of Cen A with different rigidities (energy divided by the particle's charge: $R = E/Z$)

Journal of Physics: Conference Series **531** (2014) 012003 doi:10.1088/1742-6596/531/1/012003

of 160, 80, 40 and 20 EV. Here we simulate additional and lower rigidities down to 2 EV from the center of Cen A as well as considering the northern and southern radio lobes. Another important addition is studying the effects of adding a new, detailed Kolmogorov random field component to the coherent JF12 GMF model.

Cen A was proposed to be the possible source of most cosmic rays by Farrar and Piran [8] assuming constraints on the values of turbulent extragalactic magnetic fields (See also [9]). Since the Pierre Auger Observatory released its first set of UHECR events, there have been several studies on the correlation of UHECRs with Cen A. For example, Gorbunov *et al.* discuss the possibility that Cen A could be the source of a significant part of the observed events [10]. Wibig and Wolfendale assume that all observed UHECRs were from three sources including Cen A [11]. Hardcastle *et al.* argue the possibility that the giant radio lobes of Cen A could accelerate UHECRs [12]. Other related works are described in [13–19].

The propagation method is briefly explained in Section 3 and the results of the simulation are discussed in Section 4.

2. JF12 GMF model

JF12 is a new GMF model that has been fitted to the WMAP7 Galactic synchrotron emission map [20] and more than forty thousand extragalactic rotation measures to constrain its parameters. This model includes a large scale regular component, a so-called "striated" random component that is generally arranged to follow the larger geometric features of the regular component [5], and a turbulent random field component [6]. In this study, we first consider the large scale regular component of the JF12 model. The component contains a disk field component, a toroidal halo component and an out-of-plane component as suggested by observations of external galaxies. Then, for a subset of rigidities, we add to this a turbulent random field realization characterized by a Kolmogorov power spectrum.

The r.m.s. magnitude of the random component of the JF12 model varies in space. At any point it is proportional to the magnitude of the regular component of the spiral arm structure at that location. In their study, they did not construct actual realizations of the random field due to their analysis method for constraining the field magnitudes. UHECR propagation requires such a realization so we have developed a method for propagation by creating first a small (spatial) box that contains a grid of field magnitudes which follow a Kolmogorov spectrum. Copies of this box are placed throughout the GMF volume. Each box then has its internal field magnitudes scaled according to the JF12 random component r.m.s. strength at that point. The total magnetic field is the vector sum of the regular and random field vectors. Independent random field realizations can be constructed by generating new boxes with different random number seeds.

3. Tracking

We use the latest release (v3.0.4) of *CRT* [21] which is a program to inject and propagate UHECRs from various galactic and extragalactic sources through the GMF models. *CRT* uses adaptive Runge-Kutta integration methods to determine the trajectory of a charged particle through a magnetic field according to the relativistic Lorentz force.

The method we use to track the cosmic rays through the GMF is "backtracking" particles from the Earth and recording their location and velocities when they exit the Galactic field region. Backtracking is accomplished by following the anti-particle, which simply means changing the sign of the particle's charge and reversing the incoming velocity vector. The propagation of a particle stops when it hits the pre-defined boundary of the Galaxy since the extragalactic magnetic field is assumed to have negligible magnitude in this study.

We backtrack a fixed number of cosmic rays from isotropically distributed initial locations using HEALPix [22]. HEALPix is an acronym for Hierarchical Equal Area isoLatitude

Pixelization of a sphere. In this study, we pick HEALPix resolution index 11 which defines more than 5×10^7 pixels each covering 8×10^{-4} deg^2. We backtrack one particle from the center of each pixel with a fixed rigidity R. An event with rigidity R could be a proton with energy R or an iron nucleus with energy $26R$. Several different rigidities are considered starting with $\log(R_{\rm EV})$=2 (or $R = 100$ EV) and ending at $\log(R_{\rm EV})$=0.3 (or $R \simeq 2.0$ EV) in steps of 0.05 in $\log(R)$.

UHECRs are simulated isotropically over the whole sky as described above. We identify particles of different rigidities that backtrack to regions of celestial solid angle within a radius of 3° around a chosen source location; in the first case this is the center of Cen A. The Larmor radius of a relativistic charged particle in a magnetic field B is simply given by R/B, so the use of rigidity R allows a single simulation to represent a variety of primary particle compositions with appropriately scaled total primary energies. We repeat the process for source locations centered on the tips of the northern and southern radio lobe.

4. Event Excess from Cen A

The backtracking process begins by sending (anti)particles outward in all directions. We keep for further study only those trajectories that intercept the Cen A region. This is defined by a solid angle surface with angular radius of 3° around the center of Cen A. Such a size includes a good portion of the radio galaxy which may be likely sites of acceleration and emission of cosmic rays. Since Cen A is about 3.8 Mpc away from the Earth, the 3° circle encompasses a physical distance of about 200 kpc in radius from the center, roughly midway between the central engine and the limits of the outer radio lobes which lie about 500 kpc from the center.

In the backtracking method, any particle that emerges from the volume where we have defined the Galactic magnetic field with a velocity vector pointing to the defined region around Cen A is retained for analysis. Its launch direction is then an arrival direction for a particle with opposite charge but the same rigidity accelerated at Cen A. Figure 1 shows the arrival direction distribution of the simulated events with $R = 56.23$ EV. The centroid and standard deviations of longitude and latitude of the events are shown as the black point and lines respectively. The particles are propagated through only the regular component of JF12 in this section.

The process is repeated for 35 different values of rigidity. Figure 2 shows the centroids and standard deviations of the arrival direction distributions from the center of Cen A for all simulated rigidities. Each color represents a different rigidity as indicated in the legend. These points illustrate how the arrival directions of UHECRs from the center of Cen A deviate from the source direction as a function of the rigidity of the particle. Particles with large R values deviate much less than those with smaller values. Note that for $R < 17.78$ EV, the arrival directions branch into two distinct regions on the plot; the centroid and standard deviations of such split distributions are computed and plotted separately.

An alternative hypothesis is that the northern and/or southern tips of the Cen A radio lobes are the location(s) of the source accelerator. We separately study the directions of the simulated events coming from within 3° near the tips of the northern and southern radio lobes. We find that the general stream of the events are the same, but the appearance of separated regions split away from the main thread begins at lower rigidities ($R = 12.59$ EV) for the northern tip source and at higher rigidities ($R = 22.39$ EV) for the southern tip source. The main thread is shifted towards greater latitude values for the source at the northern tip and is shifted towards smaller latitude values for the source at the southern tip, comparing to the source at the center of Cen A. These shifts are only a few degrees for the lower rigidities. The shape and orientation of the threads remain the same as for the center of Cen A source case, in the sense that lower rigidities trail towards lower longitude and latitude values. The separate regions observed for lower rigidities shift primarily in longitude by about ten degrees but maintain the same curved rigidity ordering and structure.

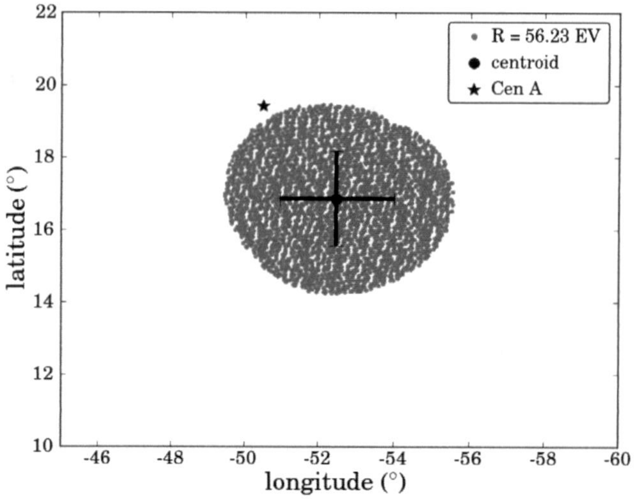

Figure 1. Directions of the simulated events with $R = 56.23$ EV originating from within 3° around the center of Cen A. The centroid and the standard deviations of these events are shown.

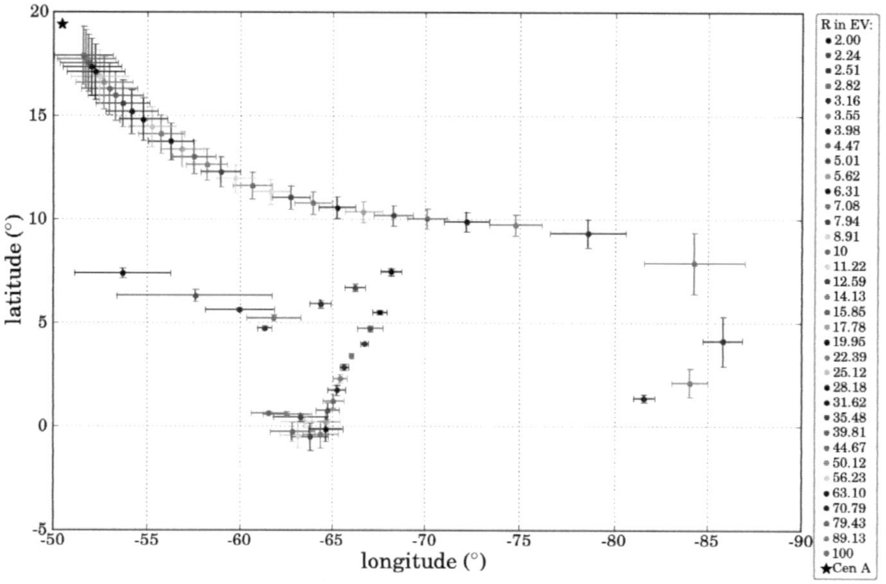

Figure 2. Centroids of the directions of the simulated events within 3° of the vicinity of the center of Cen A in Galactic coordinates. The error bars indicate the standard deviation of events with same rigidity.

The Pierre Auger Observatory has reported an excess of events (with respect to isotropic expectations) above 55 EeV within about 20° around Cen A [3]. This motivates us to examine the same quantity with our simulated arrival direction distributions.

We calculate the fraction of simulated events arriving within an angle ψ from a specified

Journal of Physics: Conference Series **531** (2014) 012003 doi:10.1088/1742-6596/531/1/012003

source direction and compare with what would be expected if arrival directions were isotropic and uncorrelated. This is done separately for the three simulation sets corresponding to the assumed Cen A source locations (north, center, or south).

To compare the Auger results, we will be interested in ultra-high energy events with energies above 55 EeV. If we assume all the particles are protons, then we will only consider simulated events with $R \geq 56.23$ EV and include all the simulated rigidities up to 100 EV in this case. On the other hand, if the events are assumed to be all iron nuclei, then the minimum simulated rigidity that we will use is $R = 2.24$, corresponding to $E \approx 58$ EeV. As very few events have ever been observed above 100 EeV, when considering iron nuclei we impose a somewhat arbitrary upper limit on the rigidities used at 7.08 EV, corresponding to a primary energy in that case of 26×7.08 EeV $= 184$ EeV.

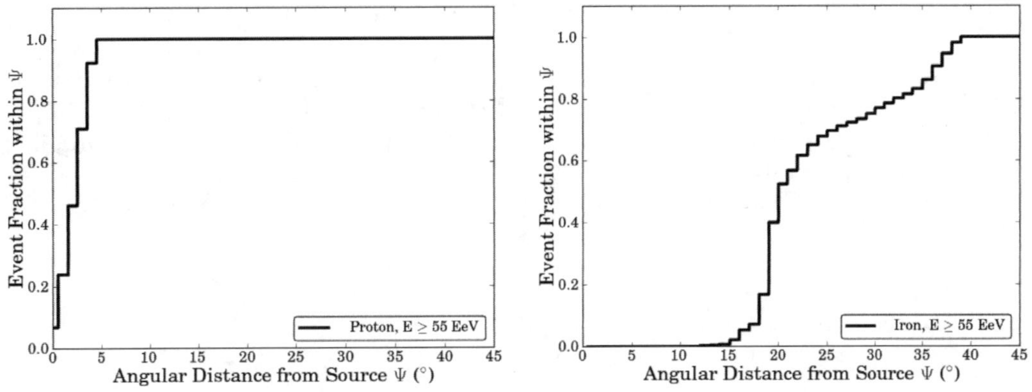

(a) Proton (b) Iron

Figure 3. Cumulative number of simulated events with $E \geq 55$ EeV as a function of angular distance from the direction of the center of Cen A for (a) protons and (b) iron nuclei.

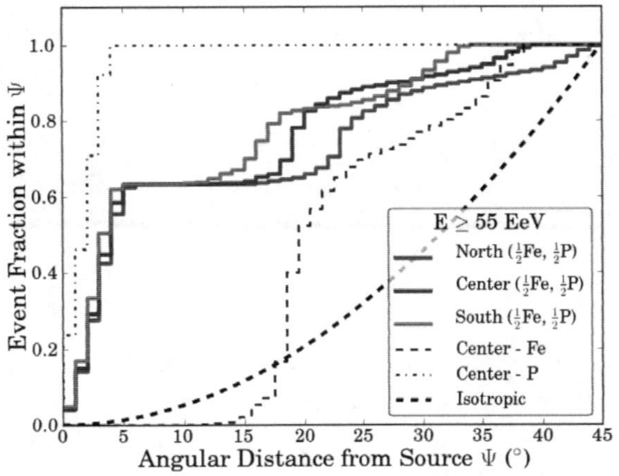

Figure 4. The excess of simulated events near Cen A. The directions have a separation angle up to 45°. Event excess from center, north, and south of Cen A are shown. The thick dashed line shows the expected isotropic flux. The thin dashed line shows only iron events. The dot-dashed line shows only proton events.

We plot the normalized cumulative number of protons and iron nuclei with $E \geq 55$ EeV as a function of angular distance from the direction of Cen A's center in Figure 3. There were no simulated events with arrival directions more than 45° away from the direction of Cen A. The simulated events are weighted by the Auger exposure [23] and an $E^{-2.6}$ differential energy spectrum [24] to match better our simulations to the observations. On average, the magnetic deflections of protons are smaller than those for iron nuclei as expected. Figure 4 shows that there is an excess above the value expected from isotropy both for protons and iron nuclei. This figure also shows that there is a visible excess above the isotropic expectation when combining the two compositions together (50% contributions from each).

We repeat this analysis for events within 3° of the northern and southern tips of Cen A radio lobes. The resulting plot is shown in Figure 4 with different colors. This shows that the excess from Cen A exists regardless of the position of the source along the large radio galaxy, although the excesses are shifted with respect to distance from the source particularly for iron nuclei.

It is evident in all cases that, when considering only the regular component of the GMF, the simulated arrival directions of primary iron nuclei match observations much better than do those from proton primaries. We note that although the angular distance distribution of events looks reasonable (Figures 3 and 4), the Auger excess within about 20° is rather isotropically distributed and does not exhibit the stream-like distribution expected for events originating from Cen A as seen in Figure 2.

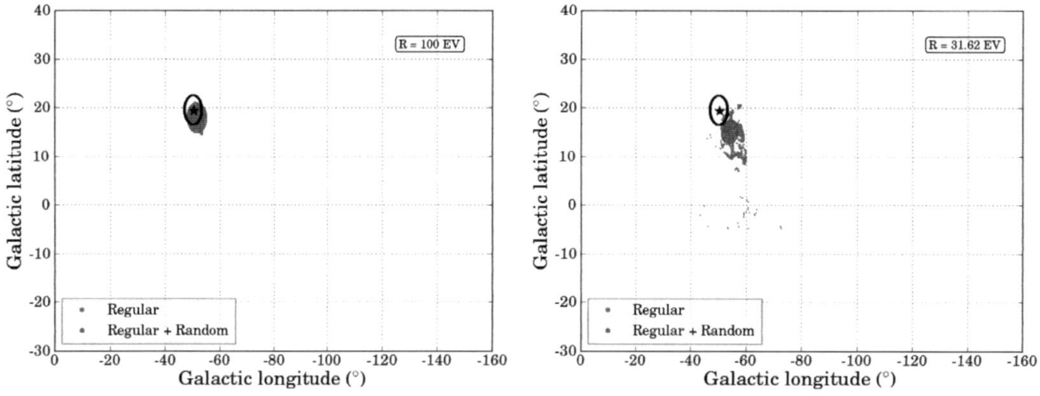

(a) $R = 100$ EV (b) $R = 31.62$ EV

Figure 5. Arrival directions of the simulated events with (a) $R = 100$ EV and (b) $R = 31.62$ EV within 3° of the vicinity of the center of Cen A in regular plus random JF12 GMF model. The blue region relates to events in regular field-only simulation. The black circle shows 3° angular distance from Cen A.

5. UHECR Deflections in JF12 GMF Model including Random Fields

Here we discuss the effects of adding the detailed Kolmogorov random field component (as discussed above in Section 2) to the JF12 regular component. The arrival directions of simulations from the center of Cen A in the JF12 regular component and the JF12 regular-plus-random components are compared in Figures 5 and 6 for $\log(R_{\mathrm{EV}}) = 2.0$, 1.5, and 1.0, respectively. Figure 6 also compares two independent realizations of the random field to confirm the consistency of the results. We see that the results of propagation do not change significantly at high rigidities, such as 100 EV and 31.62 EV, between the regular-only and regular-plus-random models. At these rigidities, evaluating events using their average angular position and

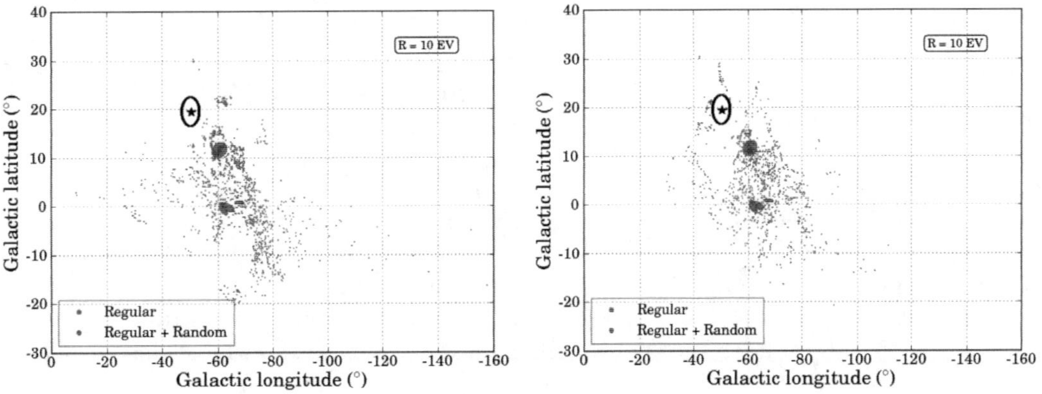

(a) Realization 1 (b) Realization 2

Figure 6. Arrival directions of the simulated events with $R = 10$ EV within $3°$ of the vicinity of the center of Cen A in the regular field plus (a) the first and (b) the second realization of the Kolmogorov random field. The blue region relates to events in regular field-only simulation. The black circle shows $3°$ angular distance from Cen A.

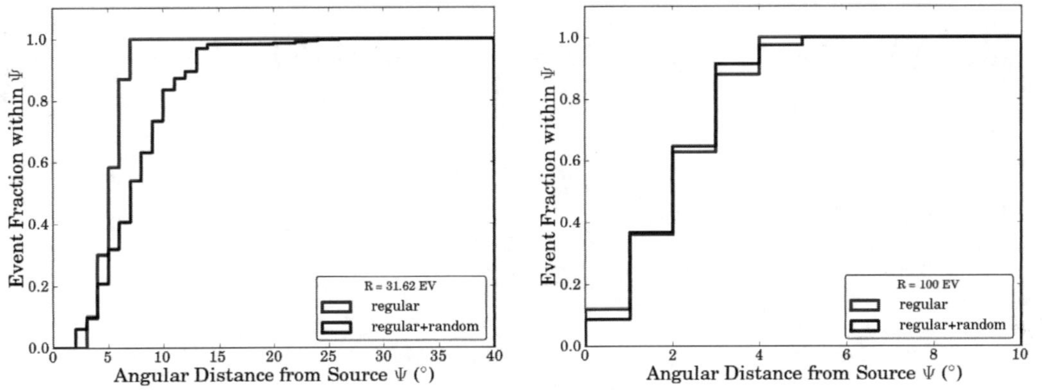

(a) R = 31.62 EV (b) R = 100 EV

Figure 7. Cumulative number of events as a function of angular distance from the direction of the center of Cen A in two cases of regular component and regular-plus-random components of JF12 GMF model. Each figure is for a different rigidity.

standard deviation still works and can be applied for study of the event excesses. At lower rigidities, such as 10 EV, the centroids are separated by a few degrees, but most importantly they are significantly smeared so that studies using just the centroid and standard deviation cannot capture the actual distribution.

The effect of including the Kolmogorov random field as we have done will change the integrated angular distributions shown earlier (Figures 3 and 4). Figure 7 illustrates the effect at two rigidities. At $R = 31.62$ EV, there is a modest but noticeable increase in the width of the integrated angular deviations, whilst at $R = 100$ EV, this effect is negligible. The impact of this will be quite different for the different assumed cases of protons or iron. For protons with energies above 55 EeV, the cumulative histograms (e.g., Figure 4) will show little change.

Journal of Physics: Conference Series **531** (2014) 012003 doi:10.1088/1742-6596/531/1/012003

But iron primaries in the same energy regime have much lower rigidity (as described above, the maximum rigidity used for iron is only $R = 7.08$ EV). It appears that the angular deviation distribution for iron nuclei will widen so much that it will be difficult to resolve it from isotropy. Neither of these cases will resemble observations as well as the outcome of including only the regular JF12 component.

Acknowledgments

Resources supporting this work were provided by the NASA High-End Computing (HEC) Program through the NASA Advanced Supercomputing (NAS) Division at Ames Research Center, consisting of time on the Pleiades supercomputing cluster awarded to Glennys Farrar. We thank Jonathan Roberts and Ronnie Jansson for their collaboration in the initial phase of this work, as well as our many colleagues in the Pierre Auger Collaboration for their many interactions and input. J.M., M.S., and A.K. acknowledge support from the Department of Energy under grants DE-0009926 and DE-FG02-91-EF0617. G.F. acknowledges support from the National Science Foundation and NASA under grants NSF-PHY-1212538 and NNX10AC96G.

References

[1] Zatsepin G T and Kuz'min V A 1966 *Sov. J. Exp. Theor. Phys. Lett.* **4** 78
[2] Greisen K 1966 *Phys. Rev. Lett.* **16** 748
[3] The Pierre Auger Collaboration 2010 *Astropart. Phys.* **34** 314
[4] The Pierre Auger Collaboration 2009 *Astropart. Phys.* **31** 399
[5] Jansson R and Farrar G R 2012 *Astrophys. J.* **757** 14
[6] Jansson R and Farrar G R 2012 *Astrophys. J.* **761** L11
[7] Farrar G R *et al.* 2013 *JCAP* **01** 023F
[8] Farrar G R and Piran T 2000 *ArXiv: astro-ph/0010370*
[9] Ginzburg V L and Syrovatskii S L 1963 *Soviet Astronomy* **7** 357
[10] Gorbunov D S *et al.* 2008 *ArXiv: astro-ph/0804.1088*
[11] Wibig T and Wolfendale A W 2009 Ôpen Astron. J. **2** 95
[12] Hardcastle M J *et al.* *MNRAS* **393** 1041
[13] Kachelrieß M *et al.* 2009 *New Journal of Physics* **11** 065017
[14] Rachen J P 2008 *XXth Rencontres de Blois, Challenges in Particle Astrophysics*
[15] Fargion D 2008 *Physica Scripta* **78** 045901
[16] Fargion D 2009 *ArXiv: 0908.2650*
[17] Rieger F M and Aharonian F A 2009 *Astron. & Astrophys.* **506** L41
[18] Moskalenko I V *et al.* 2009 *Astrophys. J.* **693** 1261
[19] Nagar N M and Matulich J 2010 *Astron. & Astrophys.* **523** A49
[20] Gold B *et al.* 2011 *Astrophys. J.* **192** 15
[21] Sutherland M *et al.* 2010 *Astropart. Phys.* **34** 198
[22] HEALPix, http://healpix.jpl.nasa.gov/
[23] Sommers P 2001 *Astropart. Phys.* **14** 271
[24] Schulz A for the Pierre Auger Collaboration 2013 *Proc. of the 33rd Int. Cosmic Ray Conf.* arXiv:1307.5059

Journal of Physics: Conference Series **531** (2014) 012004 doi:10.1088/1742-6596/531/1/012004

Sensitivity of orbiting JEM-EUSO to large-scale cosmic-ray anisotropies

Peter B. Denton[1], Luis A. Anchordoqui[2], Andreas A. Berlind[1], Matthew Richardson[1], and Thomas J. Weiler[1], for the JEM-EUSO Collaboration.

[1] Department of Physics & Astronomy, Vanderbilt University, Nashville, TN, 37235 USA
[2] Physics Department, University of Wisconsin-Milwaukee, Milwaukee, WI, 53201 USA

Speaker's E-mail: `peterbd1@gmail.com`

Abstract. The two main advantages of space-based observation of extreme-energy ($\gtrsim 10^{19}$ eV) cosmic-rays (EECRs) over ground-based observatories are the increased field of view, and the all-sky coverage with nearly uniform systematics of an orbiting observatory. The former guarantees increased statistics, whereas the latter enables a partitioning of the sky into spherical harmonics. We have begun an investigation, using the spherical harmonic technique, of the reach of JEM-EUSO into potential anisotropies in the extreme-energy cosmic-ray sky-map. The technique is explained here, and simulations are presented. The discovery of anisotropies would help to identify the long-sought origin of EECRs.

1. Introduction

The Extreme Universe Space Observatory (EUSO) is a consortium of 250 Ph.D. researchers from 27 institutions, spanning 14 countries. It is a down-looking telescope optimized for near-ultraviolet fluorescence produced by extended air showers in the atmosphere of the Earth. EUSO is proposed to occupy the Japanese Experiment Module (JEM) on the International Space Station (ISS), and collect up to 1000 cosmic ray (CR) events at and above 55 EeV (1 EeV $= 10^{18}$ eV) over a 5 year lifetime, far surpassing the reach of any ground-based project.

JEM-EUSO brings two new, major advantages to the search for the origins of EECRs. One advantage is the large field of view (FOV), attainable only with a space-based observatory. With a 60° opening angle for the telescope, the down-pointing ("nadir") FOV is

$$\pi(h_{\rm ISS}\tan(30^\circ))^2 \approx h_{\rm ISS}^2 \approx 150{,}000\,{\rm km}^2\,. \tag{1}$$

PAO has a FOV of 3,000 km^2. Thus the JEM-EUSO FOV, given in Eq. (1), is 50 times larger for instantaneous measurements (e.g., for observing transient sources). Multiplying the JEM-EUSO event rate by an expected 18% duty cycle, we arrive at a time-averaged nine-fold increase in acceptance for JEM-EUSO compared to PAO, at energies where the JEM-EUSO efficiency has peaked (at and above ~ 100 EeV). Tilting the telescope turns the circular FOV given in Eq. (1) into a larger elliptical FOV. The price paid for "tilt mode" is an increase in the threshold energy of the experiment.

The second advantage is the coverage of the full sky (4π steradians) with nearly constant systematic errors on the energy and angle resolution, again attainable only with a space-based

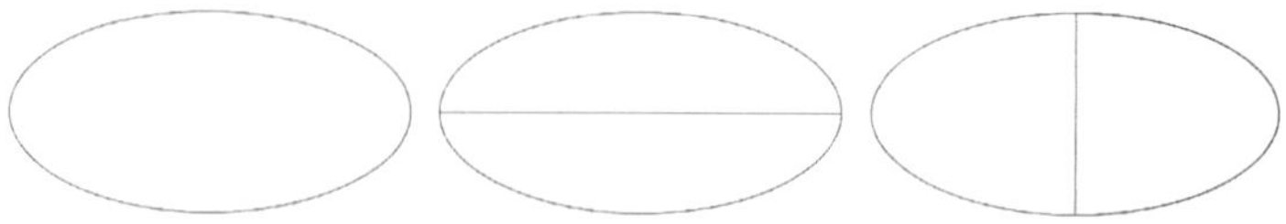

Figure 1. Nodal lines separating surplus and deficit regions of sky, for (left) $\ell = 0$, $m = 0$ monopole, and $\ell = 1$, $m = 0$ (middle) and $m = 1$ (right) dipoles.

observatory. (Combined data from ground-based observatories in the Northern and Southern hemispheres may offer full-sky coverage, but not uniformity of systematics.) This talk pursues all-sky studies of possible spatial anisotropies. The reach benefits from the 4π sky coverage, but also from the increased statistics resulting from the greater FOV. A longer study will soon be completed and published [1].

In addition to the two advantages of space-based observation just listed, a third feature provided by a space-based mission may turn out to be significant. It is the increased acceptance for Earth-skimming neutrinos when the skimming chord transits ocean rather than land. On this latter topic, just one study has been published [2]. The study concludes that an order of magnitude larger acceptance results for Earth-skimming events transiting ocean compared to transiting land. Ground-based observatories will not realize this benefit, since they cannot view ocean chords.

2. All-sky coverage and anisotropy

As emphasized by Sommers over a dozen years ago [3], an all-sky survey offers a rigorous expansion in spherical harmonics, of the normalized spatial event distribution $I(\Omega)$, where Ω denotes the solid angle parameterized by the pair of latitude (θ) and longitude (ϕ) angles:

$$I(\Omega) \equiv \frac{N(\Omega)}{\int d\Omega\, N(\Omega)} = \sum_{\ell=0}^{\infty} \sum_{|m|\leq l} a_{\ell m}\, Y_{\ell m}(\Omega)\,, \tag{2}$$

i.e., the set $\{Y_{\ell m}\}$ is complete. Averaging the $a^2_{\ell m}$ over the $(2\ell+1)$ values of m defines the rotationally-invariant "power spectrum" in the single variable ℓ, $C(\ell) = \frac{1}{2\ell+1}\sum_{|m|\leq\ell} a^2_{\ell m}$. The set $\{Y_{\ell m}\}$ is also orthogonal, obeying

$$\int d\Omega\; Y_{\ell_1 m_1}(\Omega)\, Y_{\ell_2 m_2}(\Omega) = \delta_{\ell_1 \ell_2}\, \delta_{m_1 m_2}\,. \tag{3}$$

We are interested in the real valued $Y_{\ell m}$'s, defined as $P^\ell_m(x)(\sqrt{2}\cos(m\phi))$ for positive m, $P^\ell_{|m|}(x)(\sqrt{2}\sin(|m|\phi))$ for negative m, and $P_\ell(x)$ for $m = 0$. Here, P^ℓ_m is the associated Legendre polynomial, $P_\ell = P^\ell_{m=0}$ is the regular Legendre polynomial, and $x \equiv \cos\theta$.

The lowest multipole is the $\ell = 0$ monopole, equal to the average all-sky flux. The higher multipoles ($\ell \geq 1$) and their amplitudes $a_{\ell m}$ correspond to anisotropies. Guaranteed by the orthogonality of the $Y_{\ell m}$'s, the higher multipoles when integrated over the whole sky equate to zero.

A nonzero m corresponds to $2\,|m|$ longitudinal "slices" ($|m|$ nodal meridians). There are $\ell+1-|m|$ latitudinal "zones" ($\ell-|m|$ nodal latitudes). In Figs. (1-3) we show the partitioning described by some low-multipole moments. The configurations with $(\ell, -|m|)$ are related to those with $(\ell, +|m|)$ by a longitudinal phase advance $\phi \to \phi + \frac{\pi}{2}$, or $\cos\phi \to \sin\phi$.

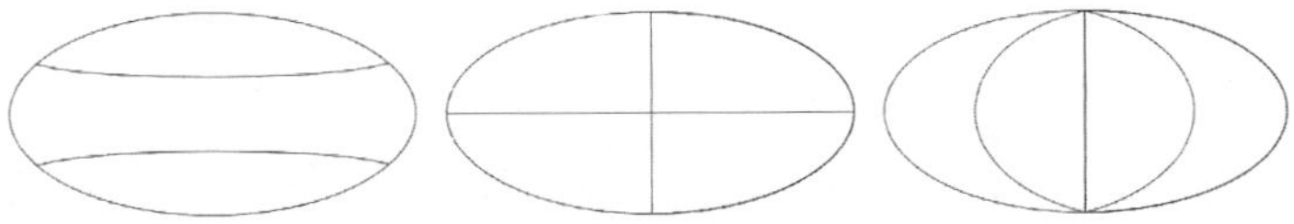

Figure 2. Nodal lines separating surplus/deficit regions of sky, for $\ell = 2$, $m = 0, 1, 2$ quadrupoles, respectively.

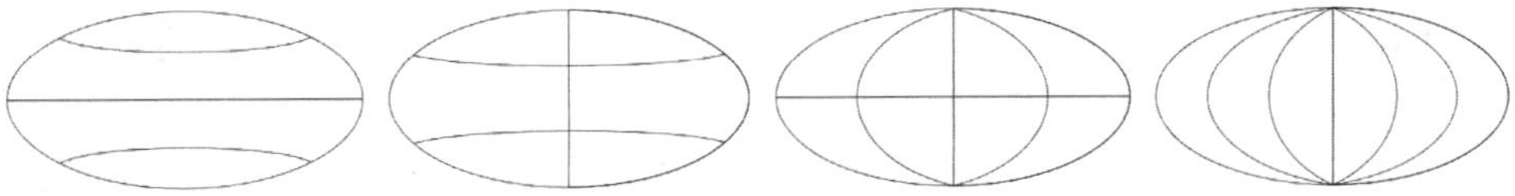

Figure 3. Nodal lines separating surplus/deficit regions of sky, for $\ell = 3$, $m = 0, 1, 2, 3$, respectively.

3. Previous anisotropy searches

The first full-sky large anisotropy search was based on the combined northern and southern hemisphere data from the SUGAR and AGASA experiments taken during a 10 yr period. Nearly uniform exposure to the entire sky resulted. No significant deviation from isotropy was seen, even at energies beyond 4×10^{19} eV [4]. More recently, the Pierre Auger Collaboration carried out various searches for large scale anisotropies in the distribution of arrival directions of cosmic rays above 10^{18} eV [5, 6].

The latest Auger study was performed as a function of both declination and right ascension in several energy ranges above 10^{18} eV, and reported in terms of dipole and quadrupole amplitudes. Again no significant deviation from isotropy was revealed. Assuming that any cosmic ray anisotropy is dominated by dipole and quadrupole moments in this energy range, the Pierre Auger Collaboration derived upper limits on their amplitudes. Such upper limits challenge an origin of cosmic rays above 10^{18} eV from non-transient galactic sources densely distributed in the galactic disk [7]. At the energies exceeding 6×10^{19} eV, however, hints for a dipole anisotropy may be emerging [8].

It must be emphasized that because previous data were so sparse at energies which will be accessible to JEM-EUSO, upper limits on anisotropy were necessarily restricted to energies below the threshold of JEM-EUSO. JEM-EUSO expects many more events at $\sim 10^{20}$ eV, allowing an enhanced anisotropy reach. In addition, JEM-EUSO events will have a higher rigidity $\mathcal{R} = E/Z$, and so will be less bent by magnetic fields; this may be helpful in identifying particular sources on the sky.

4. Comparison of all-sky JEM-EUSO to half-sky PAO

The ground-based Pierre Auger Observatory (PAO) is an excellent, ground-breaking experiment. However, in the natural progression of science, PAO will be superseded by space-based observatories. JEM-EUSO is designed to be first of its class, building on the successes of PAO.

The two main advantages of JEM-EUSO over PAO are the (i) greater FOV leading to a greater exposure at EE, and the (ii) all-sky nature of the orbiting, space-based observatory. We briefly explore the advantage of the enhanced exposure first. We consider data samples of 69, 250, 410, 680, and 1000 events. The 69 events sample is that presently published by PAO for events accumulated over three years at and above 55 EeV. The annual rate of such events at PAO is $\sim 69/3 = 23$. Thus, the 250 event sample is what PAO could attain in ten years of running.

Including the JEM-EUSO efficiency down to 55 EeV reduces the factor of 9 relative to PAO

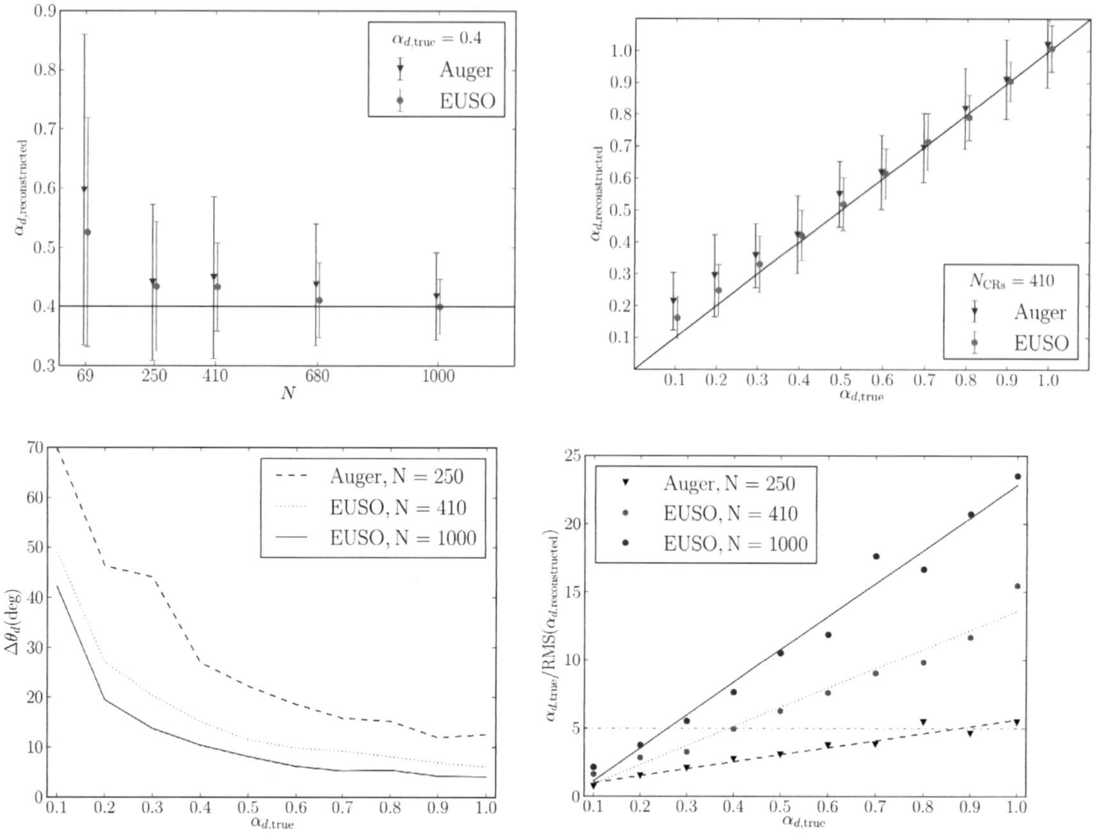

Figure 4. Reconstruction of the dipole amplitude (top panels) and angle (bottom-left panel), for PAO and JEM-EUSO. Discovery reach (bottom-right panel) of JEM-EUSO and PAO, with 5-σ horizontal line.

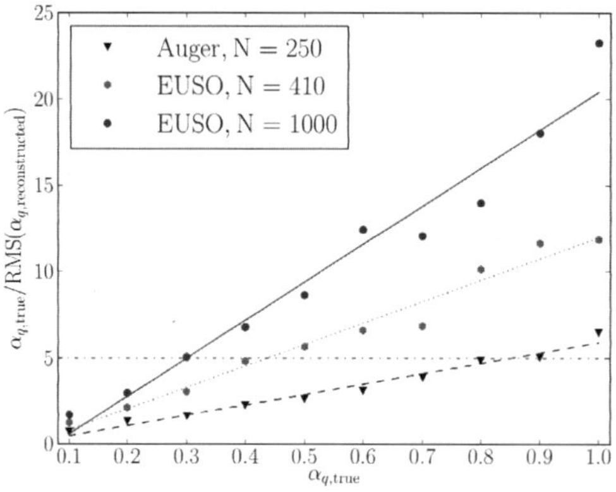

Figure 5. Discovery reach of JEM-EUSO and PAO with a 5-σ horizontal line.

down to a factor $\sim$6 at and above 55 EeV. We arrive at the 410 event sample as the JEM-

EUSO expectation at and above 55 EeV after three years running in nadir mode (or, as is under discussion, in tilt mode with an increased aperture but reduced PDM count). A 680 event sample is then expected for five years of JEM-EUSO in a combination of nadir and tilt mode. Finally, the event rate at an energy measured by High Resolution Fly's Eye (HiRes) is known to exceed that of PAO by 50%. This leads to a five-year event rate at JEM-EUSO of about 1000 events.

Now we turn to the 4π advantage. Commonly, a major component of the anisotropy is defined via a max/min directional asymmetry, $\alpha \equiv \frac{I_{\rm max}-I_{\rm min}}{I_{\rm max}+I_{\rm min}}$. For a monopole plus dipole distribution $1+\alpha_D\cos\theta$, one readily finds that $\alpha = \alpha_D$. For a monopole plus quadrupole distribution $1 - C\cos^2\theta$ (no dipole), one finds that $\alpha = \frac{C}{2-C}$, and $C = \frac{2\alpha}{1+\alpha}$.

In Fig. (4) we compare the capability of JEM-EUSO and PAO to reconstruct a dipole anisotropy. In this comparison, both advantages of JEM-EUSO, namely the increased FOV and the 4π sky coverage, are evident.

Dipole (plus monopole) plots are composed in the following way. First, we choose a dipole amplitude $\alpha_{\rm true}$ (relative to the monopole amplitude), and a dipole direction. The latter is randomly oriented, and defines the axis for the polar angle θ. Then, a given number of events, 250, 410, or 1000, are randomly distributed within the weighting $1+\alpha_{\rm true}\cos\theta$. Next, the fitting algorithm determines, as best it can, reconstructed values for α and for the dipole direction. This process is repeated 100 times, each time with a different randomly oriented dipole direction. Results are averaged, and presented in the figures. This formulation results in our observation point, the Earth, being located at the center of the dipole distribution. A dipole distribution might be indicative of a single, dominant cosmic-ray source in the presence of a magnetic field.

In the top-left panel of Fig. (4) are shown the error bars that result from a reconstruction of the dipole amplitude (we have chosen $\alpha_{\rm true} = 0.4$ for illustration), for the various N-event samples. The errors in $\Delta\alpha$ are seen to scale as $1/\sqrt{N}$ as one would expect for a Poisson distribution of events in $\cos\theta$. More significantly, the reconstruction errors in PAO for the dipole amplitude are almost twice those of JEM-EUSO, due to the limited sky-coverage of PAO (and even worse for the quadrupole, to be analyzed next) [9]. Moreover, if the dipole were aligned with the zenith angle of PAO, then PAO could easily confuse a quadrupole with a half-dipole. There is no such ambiguity with the 4π coverage of JEM-EUSO.

The top-right panel of Fig. (4) shows the error in reconstruction of the dipole amplitude, for a fixed event number $N = 410$ events. Of course, PAO will not attain an event sample of 410, but the figure correctly displays the loss of quality when the acceptance is reduced from the all-sky 4π steradians to that of PAO, even for the same number of events as JEM-EUSO. The bottom-left panel of Fig. (4) shows the reconstruction errors on the dipole direction, versus the dipole amplitude, for three event samples, PAO with 250 events, and JEM-EUSO with 410 and 1000 events. Not surprisingly, the ability to reconstruct the dipole direction improves dramatically with an increase in the dipole amplitude. For α exceeding 0.2, the 410 event JEM-EUSO sample has less that half the error of the 250 event PAO sample, and the 1000 event sample has less than a third the error of the PAO sample.

The final panel in Fig. (4) gives the number of σ for reconstruction of the dipole amplitude, as a function of the true dipole amplitude. The discriminatory power of 4π JEM-EUSO is obvious. A discovery claim (5-σ) is evident to PAO only for dipole amplitudes above 0.80. However, 410-event JEM-EUSO (3yrs) can claim discovery for an amplitude down to 0.40, and 1000-event JEM-EUSO can claim discovery all the way down to 0.28. The latter sample can reveal 3-σ "evidence" for an nonzero dipole amplitude down to 0.20.

In Fig. (5) we repeat the reconstruction comparison of JEM-EUSO and PAO, but this time for a quadrupole distribution. Our results are an average over 100 trials of a random but weighted distribution of a purely isotropic monopole and anisotropic quadrupole, the latter having a randomly chosen orientation, again with the Earth located at the center of the distribution.

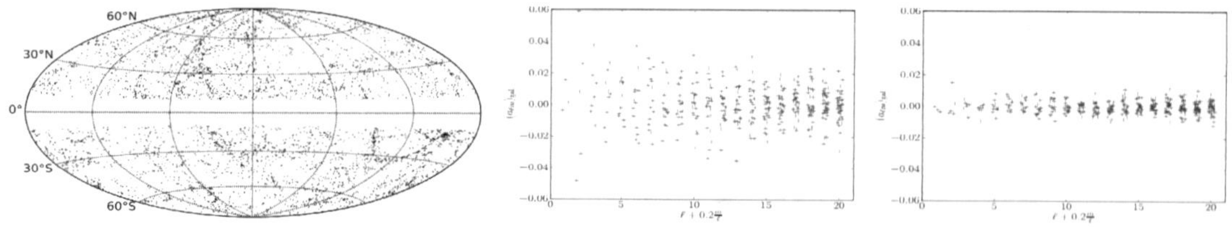

Figure 6. Sky-map of (left panel) the "5310 Galaxy-source" distribution, taken from the 2MRS catalog of galaxies [10] out to z=0.03 (115 Mpc), excluding the Galactic plane at $|b| \leq 10°$; and the fitted $a_{\ell m}$ values (center panel). (right panel) $a_{\ell m}$'s from randomly distributed 5310-event "isotropic" data, excluding the same strip of Galactic plane. The ℓ/m-dependency of the abscissa in the $a_{\ell m}$ plots is chosen [3] to increase the visibility of the $(2\ell+1)$ m values at fixed ℓ. The $|b| \leq 10°$ cut is accounted for analytically in each of the right two figures. Note the large $|a_{\ell=2}|$ values in the center figure.

There is no dipole in the input data set. A quadrupole amplitude might be indicative of a dominant Galactic distribution of sources, or even of a distribution of dominant sources in the Supergalactic Plane. A general quadrupole ($\ell = 2$) has five allowed m values. Here we consider a pure $m = 0$ distribution of events, i.e., a quadrupole with azimuthal symmetry about the quadrupole axis, as shown in the first sky-map of Fig. (2). We have again chosen $\alpha = 0.4$ for illustration. The total distribution of events mimics an oblate spheroid with relative probability distribution $1 - \frac{4}{7}\cos^2\theta$.

The same analysis as in Fig. (4) was done for the quadrupole case. A partial sky observatory fits a quadrupole much more poorly. Fig. (5) shows that JEM-EUSO can claim 5-σ discovery of a quadrupole amplitude as low as 0.4 with 410 events, and as low as 0.3 with 1000 events. Also with 1000 events, JEM-EUSO is sensitive to a 3-σ indication down to an amplitude of 0.2. On the other hand, PAO is incapable of claiming a quadrupole discovery unless the quadrupole amplitude is maximum, an unlikely value.

5. Future studies

In the near future, we will include several additional complicating, real world aspects of the spherical harmonic search for anisotropy. One relates to the energy resolution for individual events. With a spectrum falling steeply in energy, a spill-over of a lower-energy bin with isotropic events into a higher-energy bin with potentially anisotropic events will dilute the signal. While more study of this issue is warranted, this seems not to be a serious concern. PAO quotes an energy resolution $\Delta E/E$ of 22%, with 12% statistical, and about 20% systematic. Simulations in JEM-EUSO to determine the energy resolution at extreme energies are ongoing, with the present upper limit $\sim$ 30%. Also, the data sample of JEM-EUSO used in anisotropy studies is not bound to the 55 EeV threshold that PAO chose due to its limited statistics. With more statistics expected, JEM-EUSO can choose a higher-energy threshold. Of course, any anisotropy will turn on gradually in energy, and simulations must include this fact.

Other probable non-concerns are the systematic error in the angular resolution of JEM-EUSO ($\lesssim 3°$), and the bending of proton trajectories at extreme energies, given firstly by the random walk equation through extragalactic magnetic domains of strength $B_{\rm nG}$ in units of nanoGuass, and coherence size λ,

$$\delta\theta^\circ = 0.8\, Z \left(\frac{B_{\rm nG}}{E_{20}}\right) \sqrt{\frac{D\,\lambda}{10\,{\rm Mpc}^2}}\,; \tag{4}$$

where Z and E_{20} are the CR charge and energy in units of 100 EeV, respectively, and D is

Journal of Physics: Conference Series **531** (2014) 012004 doi:10.1088/1742-6596/531/1/012004

the distance traveled by the CR. One sees that for a proton ($Z = 1$) at 100 EeV, the natural unit of bending is a degree in the extragalactic magnetic field. On the other hand, heavy nuclei trajectories may be so severely bent as to eliminate even large-scale event anisotropies. So our hope hangs on protons being dominant at extreme energies. Complicating the issue is that the bending may be more or less if filamentary structure or voids are encountered en route. Subsequent to the transit of extragalactic space, the CR encounters the Galactic magnetic field, which is larger than than the extragalactic field, but is also known more precisely.

Our near-future studies will incorporate energy resolution effects, realistic estimates of Galactic and extragalactic magnetic fields, and "GZK" energy losses on cosmic radiation fields. To incorporate these effects, we will add energy and direction assignments for the individual simulated events in the context of two models, which we call the "Galaxy source model" and the "Single source model", the latter being motivated by weak evidence for a UHECR excess in the direction of Centaurus A. For the Galaxy source model, the simulated source data is weighted to the 2MRS all-sky catalog of galaxies out to $z = 0.03$ (about 115 Mpc) [10]. For the single source model, the simulated source data is fixed to a single source on the sky. For each data set, we will propagate the CRs to Earth and perform a multipole analysis of the resulting sky map.

The flavor of our work in progress can be gleaned from Fig. (6). The first panel in Fig. (6) presents the sky-map of the 5310 galaxies present in the 2MRS survey, which reaches out to z=0.03, corresponding to a distance $\sim z/H_0 \sim$ 115 Mpc. In the middle panel are shown the $a_{\ell m}$'s that result from the Galaxy-source model. As a control, the right panel displays the $a_{\ell m}$'s that result from an isotropic distribution of 5310 events with the same y scale. In both analyses, the Galactic Plane at declination below $|b| \leq 10°$ is omitted. An eyeball comparison of the two panels shows the power of the $a_{\ell m}$'s to reveal anisotropy. In particular we recover the predicted quadrupole structure in that the largest $|a_{\ell m}|$ are in the $\ell = 2$ column.

6. Conclusions

The two main advantages of space-based observation of EECRs over ground-based observatories are increased FOV and 4π sky coverage with uniform systematics. The former guarantees increased statistics, whereas the latter enables a partitioning of the sky into spherical harmonics. We have begun an investigation, using the spherical harmonic technique, of the reach of JEM-EUSO into potential anisotropies in the EECR sky-map. The discovery of anisotropies would help to identify the long-sought source(s) of EECRs.

Acknowledgments

This work is supported in part by a Vanderbilt Discovery Grant (TJW and PBD), NSF CAREER PHY-1053663 and NASA 11-APRA11-0058 Awards (LAA), Alfred P. Sloan Foundation (AAB), and NSF GAANN fellowship (MR).

References

[1] Denton P B, Anchordoqui L A, Berlind A A, Richardson M and Weiler T J in progress
[2] Palomares-Ruiz S, Irimia A and Weiler T J 2006 *Phys.Rev.* **D73** 083003 (*Preprint* astro-ph/0512231)
[3] Sommers P 2001 *Astropart.Phys.* **14** 271–286 (*Preprint* astro-ph/0004016)
[4] Anchordoqui L A, Hojvat C, McCauley T P, Paul T C, Reucroft S *et al.* 2003 *Phys.Rev.* **D68** 083004 (*Preprint* astro-ph/0305158)
[5] Abreu P *et al.* (Pierre Auger Collaboration) 2011 *Astropart.Phys.* **34** 627–639 (*Preprint* 1103.2721)
[6] Abreu P *et al.* (Pierre Auger Collaboration) 2012 *Astrophys.J.Suppl.* **203** 34 (*Preprint* 1210.3736)
[7] Abreu P *et al.* (Pierre Auger Collaboration) 2012 *Astrophys.J.* **762** L13 (*Preprint* 1212.3083)
[8] Anchordoqui L A, Goldberg H and Weiler T J 2011 *Phys.Rev.* **D84** 067301 (*Preprint* 1103.0536)
[9] There are some mathematical techniques that help PAO reconstruct, from partial sky coverage, any dipole anisotropy (see [11]). These techniques are ill-suited to quadrupole and higher multipoles.
[10] Huchra J P, Macri L M, Masters K L, Jarrett T H, Berlind P *et al.* 2011 (*Preprint* 1108.0669)
[11] Aublin J and Parizot E 2005 *Astron.Astrophys.* (*Preprint* astro-ph/0504575)

Journal of Physics: Conference Series **531** (2014) 012005 doi:10.1088/1742-6596/531/1/012005

The Loop I Superbubble and the Local Interstellar Magnetic Field

Priscilla Chapman Frisch[1]

Abstract

Recent data on the interstellar magnetic field in the low density nearby interstellar medium suggest a new perspective for understanding interstellar clouds within 40 pc. The directions of the local interstellar magnetic field found from measurements of optically polarized starlight and the very local field found from the Ribbon of energetic neutral atoms discovered by IBEX nearly agree. The geometrical relation between the local magnetic field, the positions and kinematics of local interstellar clouds, and the Loop I S1 superbubble, suggest that the Sun is located in the boundary of this evolved superbubble. The quasi-perpendicular angle between the bulk kinematics and magnetic field of the local ISM indicates that a complete picture of low density interstellar clouds needs to include information on the interstellar magnetic field.

1 Introduction

Over large spatial scales, the "ordered" versus "random" components of the interstellar magnetic field trace fundamentally different histories of the interstellar medium (ISM). The large-scale ordered interstellar magnetic field is traced through the interarm region of the galaxy with Faraday rotation measures (Rand & Kulkarni, 1989; Han, 2009) and polarized starlight (Mathewson & Ford, 1970), and is directed toward $\ell \sim 88°$. Superimposed on that pattern are "random" distortions of the large-scale magnetic field, many caused by expanding superbubbles sweeping up the ISMF during stellar evolution in spiral arms and spurs (Vallee, 1993; Elmegreen, 1993).

The Sun is located away from the density maximum of the Orion spiral arm, also known as the Local Arm, or the Orion Spur, where active star formation is ongoing. The older clusters of the Scorpius-Centaurus Association (ScoOB2) are about 120 pc away (Elias et al., 2009). The Local Bubble region around the Sun, where average interstellar densities are very low, blends into the Loop I superbubble that is the result of stellar evolution in ScoOB2 over the past 15 Myrs (Fig. 1, Frisch, 1995; Maíz-Apellániz, 2001).

The Loop I superbubble is a local example of the "random" ISMF component. The spatial scale of Loop I, with a radius of $\sim 80°$ and centered ~ 100 pc away (Heiles, 1998; Wolleben, 2007), is intermediary between the large-scale ordered ISMF component and the tiny-scale ISMF around the heliosphere. The Loop I ISMF is defined by synchrotron emission associated with the shell of the Loop I superbubble (Berkhuijsen, 1971; Wolleben, 2007). The most intense emission arises from a limb of Loop I called the North Polar Spur, which overlaps the spatial region between RA=16H and RA=20H. In this part of the sky, optical

[1]University of Chicago, Dept. Astronomy and Astrophysics, Chicago, IL 60637

Journal of Physics: Conference Series **531** (2014) 012005 doi:10.1088/1742-6596/531/1/012005

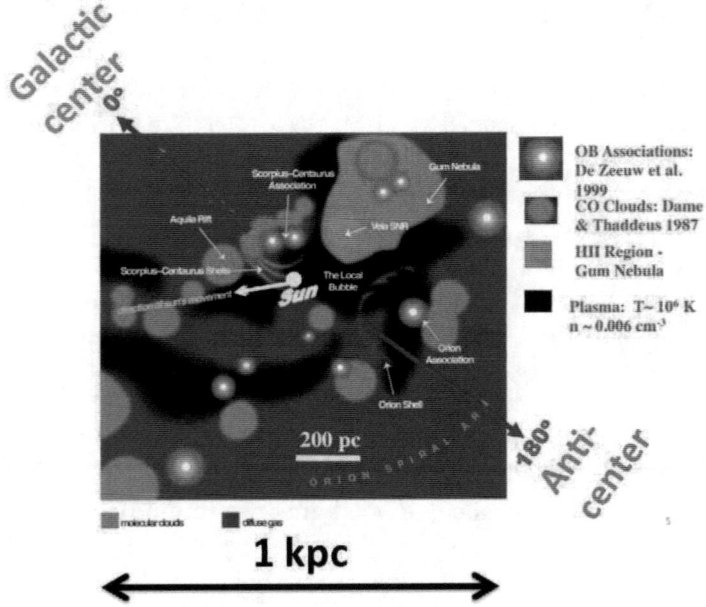

Figure 1: The distribution of molecular clouds (orange) traced by the CO 12 cm line is shown for regions within $\sim$ 1 kpc of the Sun. Stellar associations (bright starry dots) form from these molecular clouds, generating superbubble shells that compress interstellar dust and gas during expansion (gray and red shell-like arcs). The solar motion through the local standard of rest (yellow arrow) has carried the Sun through the low density interior of the Local Bubble for the past several million years. (Figure from Frisch, 2000).

polarization data show that the ISMF extends from within 10 pc of the Sun, to out beyond 100 pc (Fig. 2; Bailey et al., 2010; Santos et al., 2011). Beyond $\sim$ 100 pc, H° filaments in the North Polar Spur direction bound Loop I. The ratio of magnetic-to-thermal pressures in these filaments is $\sim$ 67 (Heiles, 1989).

At the opposite end of the spatial domain, the heliosphere samples the very local interstellar magnetic field over spatial scales of $10^2 - 10^4$ AU, where the bow wave and heliotail are included (Zank et al., 2013). The ISMF is traced by a 'Ribbon' of energetic neutral atoms (ENAs) with energies of 0.2–6 keV (McComas et al., 2009; Schwadron et al., 2009). Although the detailed physical mechanism that generates the Ribbon is not yet well-understood (McComas et al., 2011, 2014), the Ribbon forms in directions perpendicular to the ISMF as it drapes over the heliosphere (e.g. Heerikhuisen et al., 2010; Heerikhuisen & Pogorelov, 2011; Heerikhuisen et al., 2014). The best diagnostic for the direction of the ISMF that shapes the heliosphere is the center of the extraordinarily circular Ribbon arc (Funsten et al., 2013). For the 2.73 keV ENAs, the IBEX Ribbon arc is centered at $\ell = 30.5° \pm 2.6°$, $b = 57.1° \pm 1°$. The uncertainties are from the variation of the ribbon center with energy. The ratio of magnetic-to-thermal pressures in the low density weakly ionized ISM around the heliosphere is $\sim$ 1 (Frisch et al., 2011).

The low densities of nearby ISM within 10 pc, $< n > \sim 0.1$ cm^{-3} and N(H°)$< 10^{18.5}$ cm^{-2} (Wood et al., 2005), corresponds to a minimal extinction by interstellar dust of E(B-V)$\leq$

0.001 mag (Frisch et al., 2012). Interstellar material with such low densities was originally characterized as "intercloud gas" (Rogerson et al., 1973), to differentiate it from the optically vivid dense ionized gas in star-formation regions such as Orion (Fig. 1), and the cold H^o gas that dominated the 21-cm sky. Nevertheless, dust in the nearby ISM polarizes starlight and these polarizations provide the local ISMF direction within 40 pc.

Eugene Parker once asked me "what is an interstellar cloud". In the context of the very local ISM, a cloud is defined using the classical astronomical concept that a cloud is an entity if it has a velocity different from that of other gas in the same sightline. The result is that the dissection of interstellar gas into 'clouds' depends on the spectral resolution of the spectrometer. Welty has shown that the number of identified clouds increases nearly exponentially with the resolution of the spectrometer (Welty & Hobbs, 2001; Frisch, 2001). This tradition of identifying clouds by their velocity has been followed for the local ISM (e.g. Frisch et al., 2011), but the concept that clouds are individual entities is transcended by the inclusion of an interstellar magnetic field. We have found that the bulk flow of local interstellar gas past the Sun has a direction that is nearly perpendicular to the direction of the local ISMF determined from optical polarization data (Frisch et al., 2012, Section 2). This connection between cloud kinematics and magnetic field extends the concept of interstellar clouds beyond velocity components, so as to include large scale structural features in the ISMF such as the Loop I superbubble.

2 ISMF Direction from Locally Polarized Starlight

For the purpose of understanding the propagation of galactic cosmic rays (GCRs) with energies $\leq$ 10 TeV into the heliosphere vicinity, the relation between the Loop I magnetic field and the ISMF traced by the IBEX Ribbon needs to be understood. The only known way to obtain the nearby ISMF direction over scales of $\leq$ 50 pc is by extracting this information from measurements of starlight that is polarized in the local ISM (Tinbergen, 1982; Piirola, 1977; Frisch, 2010; Bailey et al., 2010; Frisch et al., 2010, 2012, 2014). Linear polarization of starlight occurs when starlight traverses an optically asymmetric interstellar medium created by charged dust grains that are aligned with respect to the ISMF (e.g., Lazarian, 2007; Hoang & Lazarian, 2014). Loop I is measured in both polarized synchrotron emission, where polarization is perpendicular to the ISMF direction, and starlight polarization, where polarization is parallel to the ISMF direction. Comparisons between the Loop I synchrotron emission (Berkhuijsen, 1971; Wolleben, 2007) and polarized starlight (Mathewson & Ford, 1970; Santos et al., 2011) show that the polarization vectors are parallel to the ISMF.

We have assembled a database of new and existing measurements of the polarizations of nearby stars with the goal of extracting the direction of the local ISMF from these data (Frisch et al., 2014, 2012, hereafter Paper III). Our method for obtaining the ISMF direction exploits the fact that the sine of the polarization position angle will be zero when the polarization vector is perfectly aligned with the direction of the ISMF, and uses that property to determines the ISMF direction that provides the best fit to the ensemble of polarization data. The stars that are included in the fit are selected based on three criteria:

(1) Stars must be within 40 pc of the Sun. Beyond that, in some directions such as the RA=16H–20H interval containing the North Polar Spur, more distant stars are sampling polarizations caused by dust in dense filaments that are not related to the low density local ISM. (2) Stars are restricted to those that are within 90° of the heliosphere nose, which is located at ℓ=5.3°, b=+12.0° (McComas et al., 2012). This restriction is partly historical because this is the spatial region where Tinbergen (1982) first discovered a nearby patch of polarizing dust. However regions more than 90° from the heliosphere nose tend to be in the anti-center hemisphere of the galaxy where less ISM is found close to the Sun (Figs. 1, 3). (3) A fully aligned dust population with no foreground depolarization will give polarization position angles that are parallel to the direction of the ISMF. Fig. 2 shows that polarizations become significantly stronger for stars beyond about 50 pc, so that grains aligned with the ISMF are unlikely to be local.

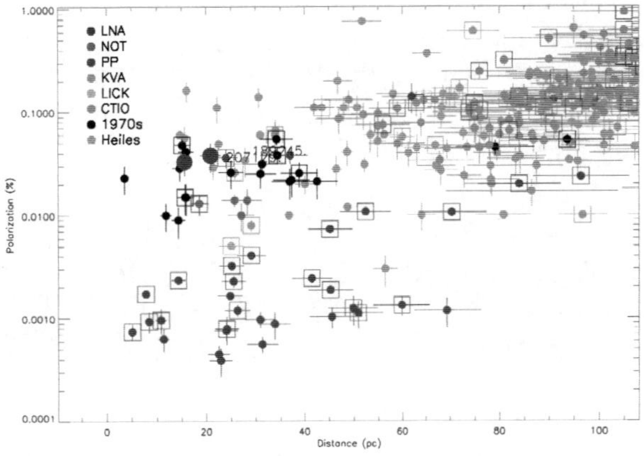

Figure 2: Strength of starlight polarizations due to the intervening ISM, for the spatial region on the sky corresponding to RA=16H–20H where polarization was found to increase with distance. The dark purple symbols represent polarizations collected in the red band where the grain extinction is less efficient. This figure is from Frisch et al. (2012), where additional details can be found.

Plots of polarization versus star distance show an upper envelope that gives the relation P(%)=9*E(B-V) for reddened stars (Serkowski, 1968). For the low color excesses of nearby stars we expect polarizations of P(%)∼0.01% or less. Special techniques are required to extract the ISMF direction from weak polarization data. Because of the low polarizations expected from the ISM within 40 pc, many stars have polarization strengths that are less than twice the mean measurement error. We therefore developed a statistical method for combining weighted polarizations so that the best-fitting ISMF direction could be derived from the polarization position angles of all of the data, including measurements where the detection is statistically insignificant (Frisch et al., 2012, 2014). The technique makes use of the statistical properties of polarization data, where the Stokes parameters Q and U are normally distributed but the polarization, $P = (\mathrm{Q}^2 + \mathrm{U}^2)^{0.5}$ is always positive (Naghizadeh-Khouei & Clarke, 1993). The directional analysis is based entirely on polarization position angles, with the polarizations themselves entering only through the statistical properties of the weighting function.

To perform the data analysis, a merit function was developed to describe the goodness-of-fit of the data to the ISMF direction. Minimization of the merit function, corresponding to minimization of the sines of the polarization position angles, then produces the best fitting ISMF direction to the data sample. We have found the direction of the local ISMF in a series of successive studies based on variations of the fitting algorithm (Frisch et al., 2010, Paper III). Each study is based on a larger dataset than the earlier studies, and used the same constraints on the scope of the data set. Using only polarization measurements with $P/dP \geq 2.0$, where P, dP are the polarization strength and mean error, an ISMF direction toward $\ell = 37° \pm 15°$, $b = 22° \pm 15°$ was found (Paper III). When the data were statistically weighted according to the Naghizadeh-Khouei & Clarke (1993) prescription above, an ISMF direction toward $\ell = 47° \pm 20°$, $b = 25° \pm 20°$ was found (Paper III). The values of the merit functions for these fits are plotted in Fig. 4. The ISMF direction based on these polarizations of nearby stars nearly agrees with the ISMF direction obtained from the IBEX Ribbon. The difference in direction is $34° \pm 28°$. One explanation is that the ISMF affecting the heliosphere and the ISMF that controls the polarizations of nearby stars are drawn from the same nearby magnetic field.

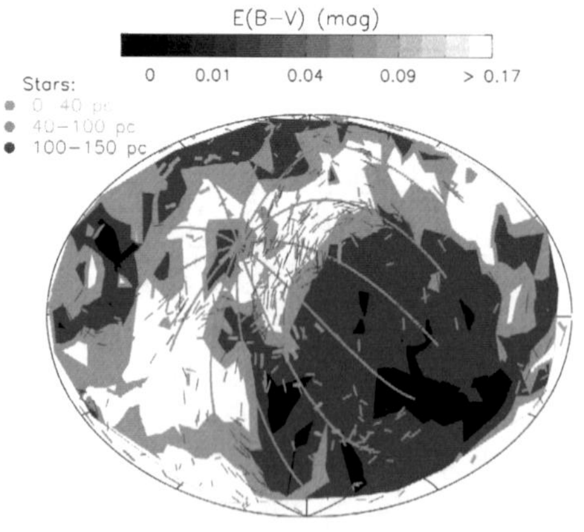

Figure 3: The color excess E(B-V) associated with the nearside of Loop I, 50–100 pc, is displayed in a reverse color scale with white regions representing the largest extinctions. Polarization vectors are color-coded by star distance. The gray-blue lines indicate field lines for the best-fitting ISMF (Fig. 4). The Loop I ISMF, traced by the purple vectors, extends to within 40 pc in directions where the polarization direction does not vary with distance. Color excesses are based the Hipparcos catalog (Perryman, 1997) and are spatially smoothed over $\sim \pm 13°$ intervals for overlapping distances. E(B-V)=0.01 corresponds to log(N(H))=19.76 cm^{-2}. Figure from Paper III.

A third paper, in preparation, refines the data sample based on the behavior of the individual polarization position angles, and obtains ISMF direction that is consistent with the previous studies and more closely agrees with the IBEX Ribbon direction (Frisch et al., 2014).

3 Local Interstellar Magnetic Field and Gas

Very young superbubble shells are semi-spherical, but the ISMF introduces a fundamental asymmetry into superbubble expansion (e.g. Ferriere et al., 1991). The small pitch angles of plasma particles that expand parallel to the ISMF will distort the spherical shape, when the ram pressure of the expanding material becomes comparable to the total pressure of the upstream ISM. The initial expansion phase of the Loop I superbubble, corresponding to times of the earliest stellar evolution in ScoOB2 about 15 Myrs ago (de Geus, 1992), would have

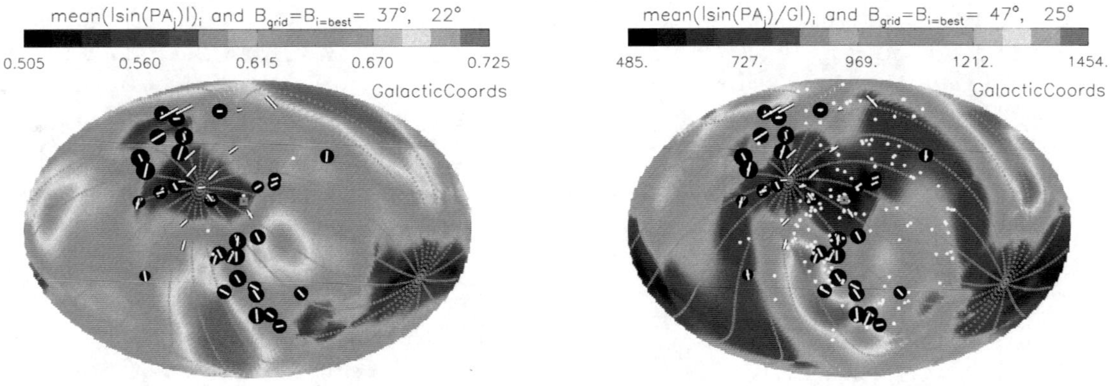

Figure 4: The value of the polarization position angle merit function evaluated over a regular grid of possible interstellar field directions is plotted, based on unweighted (left) and weighted (right) fits to the data (see Frisch et al., 2012, for more details). The symbol size is coded to increase with polarization strength, and the dots (right) represent stars where observed polarizations are not statistically significant. The gray dotted grid shows the best-fit ISMF listed in the figure title. The secondary weaker minimum centered near the heliosphere nose in the right figure (plot center) is dominated by the randomly distributed position angles contributed by low significance data points. The best-fitting ISMF is directed towards the North Polar Spur and represents the ISMF within 40 pc. The figures are centered on the galactic center with longitude increasing to the left.

expanded into an asymmetric medium resulting from the ordered ISMF field, an ISM density gradient in the galactic plane due to our location near an interarm region, and the vertical density gradient away from the galactic plane. This original star formation epoch created the Upper Centaurus subgroup of ScoOB2, and initiated hierarchal star formation that created the predecessor of the Loop I superbubble (de Geus, 1992; Crawford, 1991; Frisch, 1995; Maíz-Apellániz, 2001). The expanding superbubble would have broken out of the denser ISM beyond 100 pc where the Aquila Rift and other molecular clouds are located (Fig. 1) and expanded into the low density ISM in the Local Bubble around the heliosphere. The current geometry of the Loop I superbubble is consistent with the theoretical evolutionary models of expanding superbubbles based on fluid dynamics (Frisch, 1995). The polarization versus distance relation in Fig. 2 suggests that the local ISMF detached from the more distant ISMF of Loop I at distances of $\sim 45 - 50$ pc in the RA=16H–20H region (Fig. 2).

Wolleben (2007) has modeled the polarized radio continuum of Loop I in terms of two superbubbles, rather than a single shell. His model of the S1 shell is consistent with both the flow of ISM past the Sun and with the best-fitting ISMF direction obtained from the polarization data. The complex high-latitude structure of the ISMF associated with Loop I (Berdyugin et al., 2014), and the high-latitude H° 21-cm filaments, support the existence of two separate magnetic field structures within several hundred parsecs. The S1 shell of Wolleben is centered at $\ell = 346° \pm 5°$, $b = 3° \pm 5°$, a distance of 78 ± 10 pc away from the Sun, with a rim thickness of 19 ± 20 pc. These geometrical properties of the S1 shell were obtained from the polarized intensity expected for a spherical superbubble shell with the ISMF compressed in the rim of the shell.

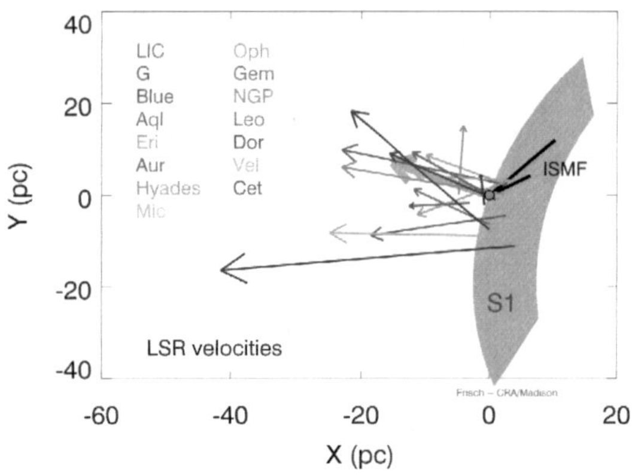

Figure 5: Velocities of interstellar clouds near the Sun are shown relative to the Loop I superbubble shell as defined by the S1 shell derived in Wolleben (2007). The velocities of the fifteen clouds derived in Redfield & Linsky (2008) are plotted in the local standard of rest, using velocities from Frisch & Schwadron (2013). The direction of the local interstellar magnetic field is the upper black line, while the IBEX ISMF direction is the lower black line. The bulk motion of the local ISM through the LSR is shown by the large thick gray arrow (Frisch et al., 2011). The local ISM flow is nearly perpendicular to the S1 shell. All quantities are projected onto the galactic plane and the Sun is located at the orange dot.

The properties of the very local ISM are fully consistent with a scenario where the local clouds are part of the rim of the S1 shell (Frisch, 2010; Frisch et al., 2010, Paper III). The cluster of local interstellar clouds (CLIC) flows past the Sun with a local standard of rest (LSR) velocity of 16.8 ± 4.6 km s^{-1}, and away from the upwind direction $\ell = 335°$, $b = -5°$. The upwind direction of the CLIC flow is within $13.6° \pm 7.1°$ of the center of the S1 shell. Fig. 5 shows the shell projected onto the galactic plane, together with the LSR bulk flow of the CLIC past the Sun (large gray arrow) and the LSR velocities of the fifteen interstellar clouds within several parsecs that have been identified by Redfield and Linsky (Redfield & Linsky, 2008; Frisch et al., 2011).

The other feature of the S1 shell that supports the association of local ISM with the S1 shell is the ISMF directions obtained from both the polarization data and IBEX Ribbon center. The angle between the CLIC bulk flow and the best-fitting ISMF to the polarization data is $76° \pm 28°$. The angle between the ISMF at the heliosphere given by the center of the ENA Ribbon, and the CLIC bulk motion, is $76.5° \pm 2.8°$. Both of these angles suggest that the local ISMF is nearly perpendicular to the bulk motion of the CLIC past the Sun. The uncertainties on the CLIC flow direction are not included for either of these estimates. Much closer, the ISMF traced by the IBEX Ribbon is perpendicular to the LSR velocity vector of the Local Interstellar Cloud that surrounds the heliosphere (Schwadron et al., 2014).

4 Continuity of ISMF over hundreds of parsecs

Salvati (2010) has derived the direction and polarity of the ISMF from four pulsars in the third galactic quadrant. Using the Faraday rotation and dispersion measure data for these pulsars, Salvati found that the ISMF direction is directed toward $\ell = 5°$, $b = 42°$, with a strength of 3.3 μG. This direction corresponds to ecliptic coordinates of $\lambda = 232°$, $\beta = 18°$, so that the ISMF would be directed upwards through the plane of the ecliptic. The ISMF direction from the pulsar data differs by 22° from the ENA Ribbon ISMF direction. Although the four pulsars are not optimally located for tracing the local ISMF since they are at distances of 160–290 pc, the agreement between the ISMF field directions found from the IBEX ENA Ribbon and pulsars suggests the presence of a nearby large scale coherent ISMF.

Together, the ISMF directions obtained from the polarization data, the ENA Ribbon ISMF data, and the pulsar data are consistent with the existence of an ISMF that is relatively non-turbulent over 150 pc or more, stretching from a pole near $\ell\sim 45°$, $b\sim 25°$ in the North Polar Spur direction, through the heliosphere location, and into the third galactic quadrant. The spatial variations in the ISMF direction over this interval are consistent with a mild curvature of the field of roughly $\sim 0.2°$ per parsec. The connectivity between this large scale ISMF and the ISMF that shapes the heliosphere affects the flux of galactic cosmic rays into the heliosphere (Schwadron et al., 2014).

Acknowledgements This work has been supported by the Interstellar Boundary Explorer mission as a part of NASA's Explorer Program. This paper is part of the proceedings of the Cosmic Ray Anisotropy workshop held September 2013 in Madison Wisconsin.

References

Bailey, J., Lucas, P. W., & Hough, J. H. 2010, MNRAS, 405: 2570.

Berdyugin, A., Piirola, V., & Teerikorpi, P. 2014, A&A, 561: A24

Berkhuijsen, E. M. 1971, A&A, 14: 359

Crawford, I. A. 1991, A&A, 247: 183

Dame, T. M., Ungerechts, H., Cohen, R. S., de Geus, E. J., Grenier, I. A., May, J., Murphy, D. C., Nyman, L. A., & Thaddeus, P. 1987, ApJ, 322: 706

de Geus, E. J. 1992, A&A, 262: 258

de Zeeuw, P. T., Hoogerwerf, R., de Bruijne, J. H. J., Brown, A. G. A., & Blaauw, A. 1999, AJ, 117: 354

Elias, F., Alfaro, E. J., & Cabrera-Caño, J. 2009, MNRAS, 397: 2

Elmegreen, B. G. 1993, in Protostars and Planets III, 97

Ferriere, K. M., Mac Low, M., & Zweibel, E. G. 1991, ApJ, 375: 239

Frisch, P. C. 1995, Space Sci. Rev., 72: 499

Frisch, P. C. 2000, American Scientist, 88: 52

Frisch, P. C. 2001, *Century of Space Science*, Eds: J. Bleeker, J. Geiss and M. C. E. Huber, Kluwer Academic Publishers", p. 647

— 2010, ApJ, 714: 1679

Frisch, P. C., Andersson, B., Berdyugin, A., Funsten, H. O., Magalhaes, M., McComas, D. J., Piirola, V., Schwadron, N. A., Slavin, J. D., & Wiktorowicz, S. J. 2010, ApJ, 724: 1473

Frisch, P. C., Andersson, B.-G., Berdyugin, A., Piirola, V., DeMajistre, R., Funsten, H. O., Magalhaes, A. M., Seriacopi, D. B., McComas, D. J., Schwadron, N. A., Slavin, J. D., & Wiktorowicz, S. J. 2012, ApJ, 760: 106.

— 2014, in preparation.

Frisch, P. C., Redfield, S., & Slavin, J. 2011, ARA&A, 49: 237

Frisch, P. C., & Schwadron, N. A. 2013, eprint = ArXiv e-prints/astro-ph:1205.4017

Funsten, H. O., DeMajistre, R., Frisch, P. C., Heerikhuijsen, J., Higdon, D. M., Janzen, P., Larsen, B., Livadiotis, G., McComas, D. J., Möbius, E., Reese, C., Reisenfeld, D. B., Schwadron, N. A., & Zirnstein, E. J. 2013, ApJ, 776: 30

Han, J. 2009, IAU Symposium, Magnetic Structure of our Galaxy, 259:455

Heerikhuisen, J., & Pogorelov, N. V. 2011, ApJ, 738: 29

Heerikhuisen, J., Pogorelov, N. V., Zank, G. P., Crew, G. B., Frisch, P. C., Funsten, H. O., Janzen, P. H., McComas, D. J., Reisenfeld, D. B., & Schwadron, N. A. 2010, ApJ, 708: L126

Heerikhuisen, J., Zirnstein, E. J., Funsten, H. O., Pogorelov, N. V., & Zank, G. P. 2014, ApJ, 784: 73

Heiles, C. 1989, ApJ, 336: 808

— 1998, ApJ, 498: 689

Hoang, T., & Lazarian, A. 2014, MNRAS, 438: 680

Lazarian, A. 2007, J. Quant. Spec. Radiat. Transf., 106, 225

Maíz-Apellániz, J. 2001, ApJ, 560: L83

Mathewson, D. S., & Ford, V. L. 1970, MmRAS, 74: 139

McComas, D. J., Alexashov, D., Bzowski, M., Fahr, H., Heerikhuisen, J., Izmodenov, V., Lee, M. A., Möbius, E., Pogorelov, N., Schwadron, N. A., & Zank, G. P. 2012, Science, 336: 1291

McComas, D. J., Allegrini, F., Bochsler, P., Bzowski, M., Christian, E. R., Crew, G. B., DeMajistre, R., Fahr, H., Fichtner, H., Frisch, P. C., Funsten, H. O., Fuselier, S. A., Gloeckler, G., Gruntman, M., Heerikhuisen, J., Izmodenov, V., Janzen, P., Knappenberger, P., Krimigis, S., Kucharek, H., Lee, M., Livadiotis, G., Livi, S., MacDowall, R. J., Mitchell, D., Möbius, E., Moore, T., Pogorelov, N. V., Reisenfeld, D., Roelof, E., Saul, L., Schwadron, N. A., Valek, P. W., Vanderspek, R., Wurz, P., & Zank, G. P. 2009, Science, 326: 959

McComas, D. J., Funsten, H. O., Fuselier, S. A., Lewis, W. S., Möbius, E., & Schwadron, N. A. 2011, Geophys. Res. Lett.: 381, L18101

McComas, D. J., Lewis, W. S., & Schwadron, N. A. 2014, Reviews of Geophysics

Naghizadeh-Khouei, J., & Clarke, D. 1993, A&A, 274: 968

Perryman, M. A. C. 1997, A&A, 323: L49

Piirola, V. 1977, A&AS, 30: 213

Rand, R. J., & Kulkarni, S. R. 1989, ApJ, 343: 760

Redfield, S., & Linsky, J. L. 2008, ApJ, 673: 283

Rogerson, J. B., York, D. G., Drake, J. F., Jenkins, E. B., Morton, D. C., & Spitzer, L. 1973, ApJ, 181: L110

Salvati, M. 2010, A&A, 513: A28

Santos, F. P., Corradi, W., & Reis, W. 2011, ApJ, 728: 104

Schwadron, N. A., Adams, F. C., Christian, E. R., Desiati, P., Frisch, P., Funsten, H. O., Jokipii, J. R., McComas, D. J., Moebius, E., & Zank, G. P. 2014, Science, 343: 988

Schwadron, N. A., Bzowski, M., Crew, G. B., Gruntman, M., Fahr, H., Fichtner, H., Frisch, P. C., Funsten, H. O., Fuselier, S., Heerikhuisen, J., Izmodenov, V., Kucharek, H., Lee, M., Livadiotis, G., McComas, D. J., Moebius, E., Moore, T., Mukherjee, J., Pogorelov, N. V., Prested, C., Reisenfeld, D., Roelof, E., & Zank, G. P. 2009, Science, 326: 966

Serkowski, K. 1968, ApJ, 154: 115

Tinbergen, J. 1982, A&A, 105: 53

Vallee, J. P. 1993, Ap&SS, 207: 109

Welty, D. E., & Hobbs, L. M. 2001, ApJS, 133: 345

Wolleben, M. 2007, ApJ, 664: 349.

Wood, B., Redfield, S., Linsky, J., Müller, H.-R., & Zank, G. 2005, ApJS, 159: 118

Zank, G. P., Heerikhuisen, J., Wood, B. E., Pogorelov, N. V., Zirnstein, E., & McComas, D. J. 2013, ApJ, 763: 20

Explanation for the Cosmic Ray Anisotropies at Small and Medium Angular Scales

G Giacinti

University of Oxford, Clarendon Laboratory, Parks Road, Oxford OX1 3PU, United Kingdom

E-mail: gwenael.giacinti@physics.ox.ac.uk

Abstract. A dipolar anisotropy in the cosmic ray (CR) arrival directions on the sky is predicted by the diffusion approximation (DA). However, the DA is not designed to predict phenomena arising on spatial scales smaller than the cosmic ray mean free path. We demonstrate here that energy-dependent smaller scale anisotropies (higher order multipoles) must naturally appear on the sky and reflect the local concrete realization of the magnetic field within the CR mean free path [1, 2]. We show that it is a viable explanation for the observed medium and small scale anisotropies. We also discuss the perspectives and priorities for the near future, and suggest that CR anisotropy measurements should become a new way of probing local interstellar and, or heliospheric fields in the future.

1. Introduction

Several experiments have detected anisotropies in the arrival directions of cosmic rays (CR) on the sky at Earth, in the $\sim$TeV–PeV energy range, see amongst others [3, 4, 5, 6, 7, 8]. Anisotropies have been found both on large (dipole) and smaller (higher order multipoles) angular scales.

The interstellar magnetic field in our Galaxy can be regarded as the sum of a regular and a turbulent field. The regular one is thought to follow spiral arms in the disk and typically varies on $\sim$kpc scales, while the turbulent one fluctuates on smaller spatial scales, from $l_{\rm max} \sim 100\,$pc down to a damping scale $l_{\rm min}$ —smaller than $\sim 1\,$AU [9]. CRs scatter on modes with fluctuation scales matching their gyroradius. As a consequence, the trajectories of TeV–PeV CRs resemble random walks. CR propagation in the interstellar medium of our Galaxy can be regarded as a diffusive process, described by the 'diffusion approximation' (DA). For more details, see References [10, 11].

The DA predicts the dipolar anisotropy, see Section 2, but cannot explain the smaller scale anisotropies (SSA). We show in Section 3 that the concrete realization of the magnetic field (interstellar and/or heliospheric depending on the CR rigidity E/Z) within $\sim$ a CR mean free path (MFP) from Earth naturally explains the presence of higher order multipoles on the sky. Indeed, the DA predictions represent an ensemble average over all possible local magnetic field configurations. Therefore, they do not reflect the exact anisotropy patterns on the sky for a specific observer. In practice, the concrete realization of magnetic fields within a CR MFP from the observer gives rise to a specific pattern of SSA, on top of a dipolar anisotropy.

This implies that observations of SSA should become in the future a new and promising way to probe our local interstellar and heliospheric fields, see Section 4.

Journal of Physics: Conference Series **531** (2014) 012006 doi:10.1088/1742-6596/531/1/012006

2. The large scale anisotropy

The DA predicts a dipolar anisotropy $\boldsymbol{\delta}$, whose direction and amplitude are related to the local gradient of CR density $\boldsymbol{\nabla} n$ by $\boldsymbol{\delta} = 3\mathcal{D}(E/Z)/c \cdot \boldsymbol{\nabla} n/n$, where $\mathcal{D}(E/Z)$ denotes the local value of the CR diffusion coefficient. References [12, 13] put constraints on the value of the diffusion coefficient. We take $\mathcal{D}(E/Z) \approx 10^{28} \left(\frac{E/Z}{3\,\mathrm{GeV}}\right)^{\gamma} \left(\frac{H}{\mathrm{kpc}}\right) \mathrm{cm}^2\,\mathrm{s}^{-1}$, with $\gamma \approx 0.45$, and $H \sim 1\,\mathrm{kpc}$ as an order of magnitude of the scale height of the (magnetized) Galactic halo. Then, if $|\boldsymbol{\nabla} n/n| \sim 1/H$, this yields $|\boldsymbol{\delta}| \sim 10^{-3} \left(\frac{E/\langle Z\rangle}{20\,\mathrm{TeV}}\right)^{\gamma}$. Such a value roughly corresponds to observations at $\sim (1-10)$ TeV energies. For detailed studies, see References [14, 15].

In practice, the anisotropy is dominated by one (or a few) recent nearby sources [15]. Also, in a given energy range, the anisotropy does not necessarily increase with energy [15]. We also remind the reader that, at TeV–PeV energies, the CR gyroradius is negligible compared to the coherence length $l_c \sim 10\,\mathrm{pc}$ of the interstellar turbulence. This implies that the dipole does *not* point in the direction of $\boldsymbol{\nabla} n/n$ [1]. The diffusion approximation predicts the *ensemble averaged* dipole. In reality, the local magnetic field around Earth has a *given* configuration —which does not strongly change on the time scales and spatial scales which are relevant here. See References [16, 17, 18] for more information on the interstellar magnetic field within a few tens of parsecs from Earth.

3. Anisotropies at medium and small angular scales

The DA only predicts a dipolar anisotropy and no higher order multipoles. Previous works have proposed a variety of mechanisms that can generate SSA under some circumstances: We refer the reader to References [19, 20, 21, 22]. More recently, Reference [23] made the interesting suggestion that SSA at $\sim (1-10)$ TeV may be due to heliospheric electric fields.

In this work, we propose that SSA are a *natural* consequence of the *given magnetic field configuration within roughly a CR mean free path* $\lambda_{\mathrm{mfp}}(E/Z)$ *from Earth.* Reference [24] supports such a view and, very interestingly, demonstrates that the observed SSA power spectrum at multi-TeV energies would be compatible with it. CRs with different rigidities E/Z scatter on different wavelengths of the local turbulence. This implies that SSA must be rigidity-dependent, which is in line with observations.

In the previous Section, we have already stressed that the DA only predicts an ensemble averaged value for the dipole. The actual amplitudes and directions of the dipoles measured by observers in different realizations of the magnetic field will differ (more or less) from these predictions, even if $\boldsymbol{\nabla} n/n$, l_c, the turbulence power spectrum and the root mean square amplitude of the field are fixed. Conceptually, this is also the reason why the DA fails to predict the higher order multipoles. From one realization of the local field to another, SSA will statistically cancel one another if averaged over a large amount of magnetic field realizations, leaving only a dipole —the ensemble averaged one, which is predicted by the DA. However, in reality, the local magnetic field has a given configuration that leads to the existence of specific SSA bearing its imprints. The interstellar magnetic field does not vary sufficiently quickly for SSA at energies $E \leq 1\,\mathrm{TeV}$ to significantly change during the life time of experiments ($\approx 10\,\mathrm{yr}$). Indeed, both the velocity of the interstellar medium and the Alfvén velocity are of the order of $10\,\mathrm{km\,s^{-1}}$, and $10\,\mathrm{km\,s^{-1}} \times 10\,\mathrm{yr} \leq r_{\mathrm{L}}(\mathrm{TeV})$, where $r_{\mathrm{L}}(\mathrm{TeV})$ is the Larmor radius of a TeV CR in a $\approx 4\,\mu\mathrm{G}$ field. Let us note that the argument presented here is valid for *any magnetic field within a cosmic ray mean free path from Earth.* As first realized by Desiati & Lazarian [25], the CR MFP should start to be of the order of the size of the heliotail for CRs with $E \leq 10\,\mathrm{TeV}$. Therefore, they proposed that, below such energies, one could start to probe the configuration of heliospheric magnetic fields —see also discussion in the next Section.

[1] See, amongst others, the talk given by Ming Zhang during this workshop.

We now verify numerically, with the code presented in Reference [26], that SSA do appear on top of the dipolar anisotropy, provided a local gradient of CR density ∇n exists. We propagate individual CRs with $E/Z = 10\,\text{PeV}$ in an isotropic turbulent magnetic field with root mean square strength $B_{\text{rms}} = 4\,\mu\text{G}$. For a detailed description of the procedure, see [1]. Using a FFT, the field is precomputed on a three-dimensional grid in physical space from its power spectrum in reciprocal space —see amongst others References [27, 28] for more details on this technique. We use here a turbulent field with a Kolmogorov spectrum ($\mathcal{P}(k) \propto k^{-5/3}$), which contains fluctuations on scales between $l_{\text{min}} = 0.2\,\text{pc}$ and $l_{\text{max}} = 150\,\text{pc}$. Magnetic field fluctuations are then resolved down to scales $\approx r_{\text{L}}(10\,\text{PeV})/10$, which ensures that the field contains modes on which CRs scatter resonantly (i.e. those with spatial fluctuation scales $\approx r_{\text{L}}$). Using anisotropic turbulence or changing the other parameters of the turbulence do not change the main conclusion of this work.

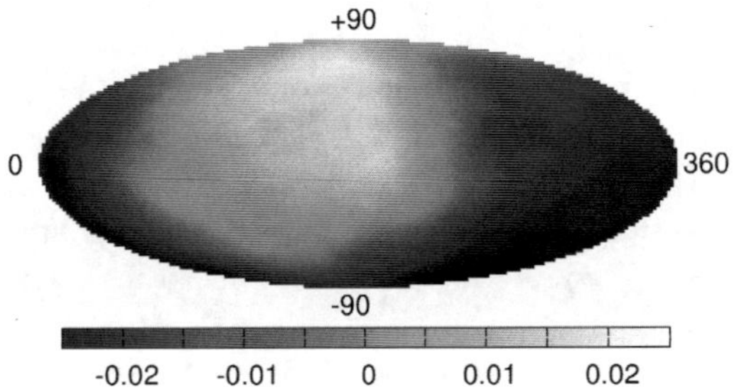

Figure 1. Simulated deviation $(\Delta N - \langle N\rangle)/\langle N\rangle$ of the CR flux on the sky at Earth from the average flux $\langle N\rangle$, after smoothing over 90° circles. Result for a given realization of the local turbulent magnetic field, see text for parameters. A large scale anisotropy can be clearly seen.

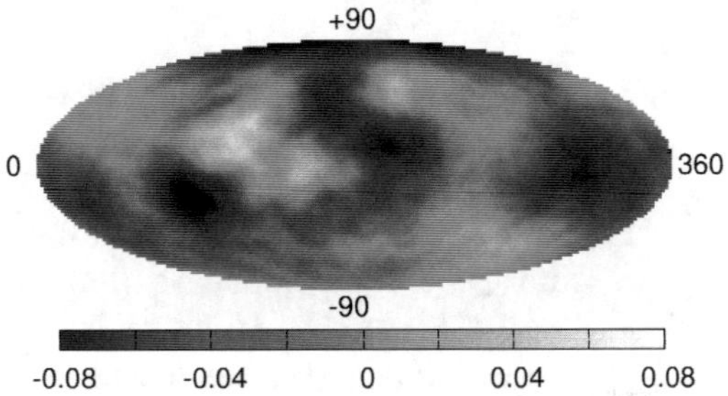

Figure 2. Same as Figure 1, after *subtracting the dipole* and smoothing over 20° circles. Same concrete realization of the turbulent magnetic field. Anisotropies at small and medium angular scales are visible, showing that higher order multipoles are present.

In order to diminish the computing time, we take a large relative gradient of CR density $|\nabla n/n| = (290\,\text{pc})^{-1}$. Figure 1 (respectively Figure 2) shows the resulting sky map for the large scale anisotropy (respectively the medium and small scale anisotropies). In Figure 1, we plot the relative deviation $(\Delta N - \langle N\rangle)/\langle N\rangle$ of the CR flux, after averaging over circles with 90° radii, from the average flux on the sky $\langle N\rangle$. For Figure 2, we have subtracted the dipolar anisotropy, and we show the remaining anisotropies at smaller scales (SSA) after averaging over circles with 20° radii.

We verified that these SSA are rigidity-dependent. In reality, experiments such as IceCube present results for broad energy ranges (and different charges), see Figure 3 of Reference [7]. SSA from CRs with different rigidities add up non-constructively. Observed SSA are then expected to have an amplitude smaller than the amplitude of the dipole. We find that for a CR distribution with a median energy $E_{\text{m}} = 10\,\text{PeV}$ and width $\Delta E/E_{\text{m}}$ set to that for the IceCube observations at $E_{\text{m}} = 20\,\text{TeV}$, the amplitude of SSA is roughly 10 times smaller than the dipole, which agrees with observations.

For a quantitative and detailed theoretical explanation for the origin of SSA, we refer the reader to our main work [1]. Very qualitatively, if CRs were travelling on *straight lines* from a 'last scattering surface' at $\lambda_{\text{mfp}}(E/Z)$ from Earth, only a dipole would be present: It would point to the direction of the local gradient of CR density ∇n. In reality, their trajectories are curved in the *given* configuration of the local magnetic field within $\approx \lambda_{\text{mfp}}(E/Z)$ from Earth,

Journal of Physics: Conference Series **531** (2014) 012006 doi:10.1088/1742-6596/531/1/012006

which reshuffles parts of the dipole to smaller patches in different directions on the sky: the SSA.

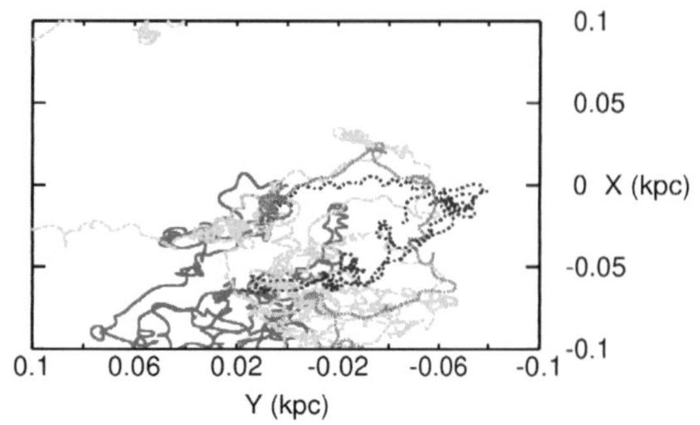

Figure 3. Trajectories of four CRs that reach Earth in four different directions on the sky: (l,b) = (180°,0°) —red line, (181°,0°) —green line, (199°,0°) —blue line, and (200°,0°) —magenta line. Computations are done for $E/Z = 10\,\mathrm{PeV}$ CRs in a concrete realization of the turbulent magnetic field, see text. Trajectories are projected on a panel of lateral size 200 pc which contains the local gradient of CR density, ∇n. ∇n points towards increasing Y, and the Earth is located at (X,Y) = (0,0).

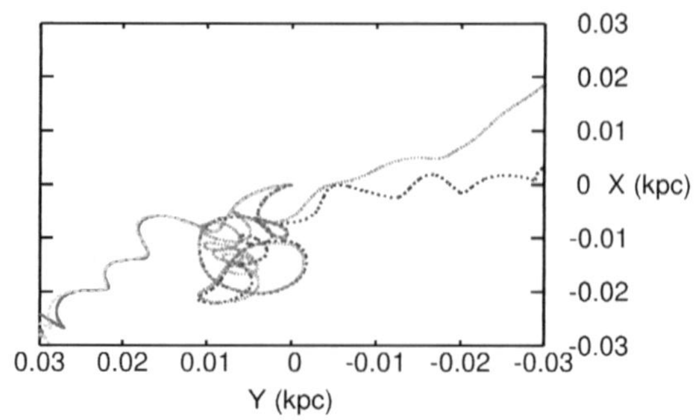

Figure 4. Zoom on Figure 3, showing CR trajectories in a panel with 60 pc lateral size. Within approximately a CR mean free path, most CRs which arrive in a SSA 'hot spot' (resp. 'cold spot') have trajectories which point back towards larger (resp. lower) n.

Figures 3 and 4 illustrate this: They show the trajectories of four CRs arriving at Earth, two being in a 'cold spot' of the SSA (i.e. a sky patch in which the CR flux is lower than expected if only a dipolar anisotropy were present), and two in a 'hot spot', see captions of the figures. In Figure 3, one can see that all four CRs may come from any direction, as should be expected. However, in Figure 4, which is a zoom to a region of size $\approx \lambda_{\mathrm{mfp}}(E/Z)$ around Earth, one can see that CRs from the cold (respectively hot) spot tend to preferentially come from the direction of lower (respectively larger) CR densities n. Therefore, the last part of the CR trajectories within $\approx \lambda_{\mathrm{mfp}}(E/Z)$ from Earth determines the given local configuration of SSA, which is in line with the theoretical model presented above.

4. Perspectives

We think that, to some extent, one now has the basic theoretical ingredients that are needed to build a self-consistent picture for the origin of the TeV-PeV CR dipole and SSA. Further CR anisotropy observations, as well as better knowledge/understanding of heliospheric fields and of local interstellar fields, should help in building such a picture.

Concerning the anisotropies at medium and small angular scales, some of the main questions which remain to be solved are :

- Below which energy do SSA start to probe heliospheric fields (as proposed in Reference [25])? Are TeV anisotropies at small scales due to the heliospheric fields?
- Is the effect of heliospheric electric fields on SSA (as proposed in Reference [23]) dominant or sub-dominant at $\sim (1-10)$ TeV energies?

- Could some simulations of heliospheric fields reproduce, or at least be somehow compatible with, the observed TeV SSA?
- What is the expected amplitude and power spectrum of SSA from the local interstellar magnetic fields, from TeV to PeV energies? And how well would it fit observations from TeV to PeV energies?
- What is the proper explanation for the anisotropy patterns observed by IceCube and IceTop at 400 TeV and 2 PeV?

In principle, one should be able to shed more light on these questions by combining the current knowledge of the local interstellar magnetic field, heliospheric fields, and checking their impact of the SSA.

We predict that, once these questions will be better understood, our field will enter a new era. Measurements of CR anisotropies on all angular scales (dipole and higher order multipoles) will enable one to deduce precious information on local magnetic fields, within the cosmic ray mean free path. In the future, this may become one of the most powerful ways to *measure such fields* (strength, structure, and maybe potential temporal variations).

5. Conclusions

We have shown that small scale anisotropies naturally arise from the *concrete local* magnetic field configuration, within $\sim$ a CR mean free path from Earth. At PeV energies, CR anisotropies probe the local interstellar turbulence within several tens of parsecs. At lower energies, below ~ 10 TeV, the same argument holds, and one may start to probe heliospheric fields [25].

We predict that the field is about to enter a new phase. In the near future, one will start to use CR anisotropy observations on different angular scales in order to gain information of the still rather poorly known local interstellar and heliospheric magnetic fields. Since the CR mean free path depends on the rigidity E/Z, CR anisotropy observations at different energies will in principle enable one to study magnetic fields at different distances from Earth, from the heliosphere to a few tens of parsecs.

Acknowledgments

The author acknowledges funding from the European Research Council under the European Community's Seventh Framework Programme (FP7/2007 2013) / ERC grant agreement No. 247039. He thanks the Committee for organizing this fascinating and fruitful workshop, and acknowledges several interesting discussions with, amongst others, Segev BenZvi, Pasquale Blasi, Paolo Desiati, Priscilla Frisch, Alex Lazarian, Mikhail Malkov, Philipp Mertsch, Nikolai Pogorelov, Marcos Santander and Nathan Schwadron.

References

[1] Giacinti G and Sigl G 2012 *Phys. Rev. Lett.* **109** 071101
[2] Giacinti G 2013 *Plasma Phys. Control. Fusion* **55** 124007
[3] Guillian G *et al.* [Super-Kamiokande Collaboration] 2007 *Phys. Rev.* D **75** 062003
[4] Amenomori M *et al.* [Tibet AS-gamma Collaboration] 2006 *Science* **314** 439
[5] Abdo A A *et al.* [Milagro Collaboration] 2008 *Phys. Rev. Lett.* **101** 221101
[6] Abdo A A *et al.* [Milagro Collaboration] 2009 *Astrophys. J.* **698** 2121
[7] Abbasi R *et al.* [IceCube Collaboration] 2012 *Astrophys. J.* **746** 33
[8] Aartsen M G *et al.* [IceCube Collaboration] 2013 *Astrophys. J.* **765** 55
[9] Han J-L, Ferriere K and Manchester R N 2004 *Astrophys. J.* **610** 820
[10] Ginzburg V L, Dogiel V A, Berezinsky V S, Bulanov S V and Ptuskin V S 1990 *Astrophysics of Cosmic Rays* (Amsterdam: North-Holland)
[11] Strong A W, Moskalenko I V and Ptuskin V S 2007 *Ann. Rev. Nucl. Part. Sci.* **57** 285
[12] Strong A W and Moskalenko I V 1998 *Astrophys. J.* **509** 212
[13] Di Bernardo G, Evoli C, Gaggero D, Grasso D and Maccione L 2010 *Astropart. Phys.* **34** 274

[14] Erlykin A D and Wolfendale A W 2006 *Astropart. Phys.* **25** 183;
[15] Blasi P and Amato E 2012 *J. Cosmol. Astropart. Phys.* **1201** 010
[16] Frisch P C, Andersson B-G, Berdyugin A, Funsten H O, Magalhaes M, McComas D J, Piirola V, Schwadron N A, Slavin J D and Wiktorowicz S J 2010 *Astrophys. J.* **724** 1473
[17] Frisch P C 2012 *AIP Conf. Proc.* **1436** 295
[18] Frisch P C, Andersson B-G, Berdyugin A, Piirola V, DeMajistre R, Funsten H O, Magalhaes A M, Seriacopi D B, McComas D J, Schwadron N A, Slavin J D and Wiktorowicz S J 2012 *Astrophys. J.* **760** 106
[19] Drury L and Aharonian F A 2008 *Astropart. Phys.* **29** 420
[20] Malkov M A, Diamond P H, Drury L and Sagdeev R Z 2010 *Astrophys. J.* **721** 750
[21] Lazarian A and Desiati P 2010 *Astrophys. J.* **722** 188
[22] Kotera K, Perez-Garcia M A and Silk J 2013 *Phys. Lett. B* **725** 196
[23] Drury L 2013 The problem of small angular scale structure in the cosmic ray anisotropy data *Preprint* arXiv:1305.6752
[24] Ahlers M 2014 *Phys. Rev. Lett.* **112** 021101.
[25] Desiati P and Lazarian A 2013 *Astrophys. J.* **762** 44
[26] Giacinti G, Kachelriess M, Semikoz D V and Sigl G 2012 *J. Cosmol. Astropart. Phys.* **1207** 031
[27] Casse F, Lemoine M and Pelletier G 2002 *Phys. Rev.* D **65** 023002
[28] De Marco D, Blasi P and Stanev T 2007 *J. Cosmol. Astropart. Phys.* **0706** 027

First Results of the AMS-02 Experiment on the ISS

Jorge Casaus, on behalf of the AMS Collaboration
Centro de Investigaciones Energéticas, Medioambientales y Tecnológicas, CIEMAT.
Av. Complutense 22, Madrid E-28040, Spain

E-mail: Jorge.Casaus@ciemat.es

Abstract. The Alpha Magnetic Spectrometer has recently released a first set of precise measurements on the cosmic rays detected from the International Space Station. The results on the positron fraction in the energy range from 0.5 to 350 GeV and the search for positron anisotropy are summarized. The very accurate data show that the positron fraction is steadily increasing from 10 to ~250 GeV. The positron fraction spectrum shows no fine structure and the positron to electron ratio shows no observable anisotropy. This is not consistent with only the secondary production of positrons. The parametrization of the positron fraction and the $e^+ + e^-$ flux measurements within the framework of a minimal model provides a complete description of the positron source contribution. The predictions for different scenarios accounting for the positron source component can thus be directly compared to this measurement.

1. Introduction

The Alpha Magnetic Spectrometer (AMS-02) [1] is a general purpose high-energy particle physics detector operating on the International Space Station (ISS) since 19 May 2011. The goal of the experiment is to carry out precise measurements of cosmic rays in the energy range from ~1 GeV to ~1 TeV. Accurate studies of the fluxes of individual components of cosmic rays are achieved thanks to the excellent particle identification and energy resolution of the detector. In particular, the measurement of antimatter in cosmic rays with unprecedented precision may have deep implications in fundamental physics.

First results of the experiment, obtained on less than a 10% of the expected AMS sample, include the measurement of the positron fraction, that is, the fraction of positron to electron plus positron fluxes up to 350 GeV [2, 3, 4], as well as preliminary measurements of individual positron and electron fluxes and positron plus electron fluxes [5, 6], proton flux to 1.8 TV [7], helium flux to 3 TV [8] and boron to carbon ratio to 670 GeV/n [9].

The excess of positrons above ~10 GeV with respect to the expectations for secondary production is one of the most striking anomalies in cosmic rays and its origin has been widely discussed, including production in standard astrophysical sources [10], dark matter annihilation in the halo of the Galaxy [11] or in the Earth's vicinity [12]. Distinctive signatures include the presence of structures in its energy spectrum, the existence of a high energy cutoff or the detection of some degree of anisotropy in their arrival direction [13]. Only precise measurements of cosmic positrons in a wide energy range may help to elucidate it.

The description of the measurement of the positron fraction in the energy range from 0.5 to 350 GeV and the systematic search for anisotropies above 16 GeV is presented. The interpretation of the results within the framework of different models will be discussed.

2. The AMS detector

The layout of the AMS detector is shown in Fig. 1. It consists of 9 planes of precision Silicon Tracker; a Transition Radiation Detector, TRD; four planes of Time of Flight counters, TOF; a Magnet; an array of anticoincidence counters, ACC, surrounding the inner Tracker; a Ring Imaging Čerenkov detector, RICH; and an Electromagnetic Calorimeter, ECAL.

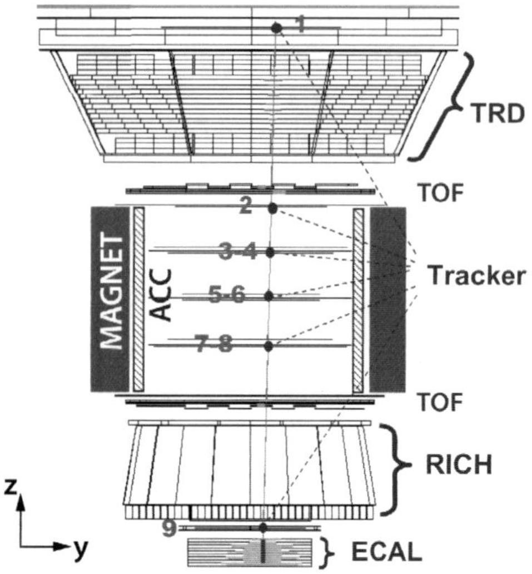

Figure 1. A 369 GeV positron event as measured by the AMS detector on the ISS in the (y-z) plane. Tracker planes 1-9 measure the particle charge and momentum. The TRD identifies the particle as an electron/positron. The TOF measures the charge and ensures that the particle is downward going. The RICH measures the charge and velocity. The ECAL independently identifies the particle as an electron/positron and measures its energy. Electrons and positrons are identified by (i) an electron signal in the TRD, (ii) an electron signal in the ECAL, and (iii) the matching of the ECAL shower energy and the momentum measured with the tracker and magnet.

There are three main detectors that allow a significant reduction of the proton background in the identification of the positron and electron samples. The TRD, which provides a proton rejection of 10^3 to 10^4 at 90% $e^{\pm}$ efficiency; the Tracker, which determines the trajectory, the absolute value of the charge and its sign, with a maximum detectable rigidity of 2 TV; and the ECAL, which provides a precise measurement of the $e^{\pm}$ energy and a proton rejection of 10^4 when combined with the energy-momentum matching requirement $E/p>0.75$.

The performance of the different subdetectors has been validated on detailed Monte Carlo simulations and test beam data. Moreover, the independent e/p separation capabilities of TRD, Tracker and ECAL allows the selection of clean electron and proton samples on ISS data. Therefore, reference distributions for the TRD estimator, formed from the ratio of the loglikelihood probability of the $e^{\pm}$ hypothesis to that of the proton hypothesis, are obtained on samples selected using ECAL and Tracker. Likewise, distributions for protons and electrons of the ECAL estimator, based on a Boosted Decision Tree, BDT, algorithm [14] constructed using the 3D shower shape in the ECAL, are obtained on samples selected with the TRD.

3. Measurement of the positron fraction

The reduction of the proton background in the identification of the positron and electron samples is achieved by means of the TRD, the ECAL and the tracker. Events are selected by requiring a track in the TRD and in the tracker, a cluster of hits in the ECAL compatible with and electromagnetic shower with an energy above the geomagnetic cutoff, and a measured velocity $\beta \sim 1$ in the TOF consistent with a downward-going $Z = 1$ particle.

The selection efficiency for positrons and electrons is estimated to be ~90% in the acceptance of the ECAL. The selected sample contains ~6,800,000 primary positrons and electrons and ~700,000 protons.

Selected events are grouped into ECAL energy bins. In every energy bin, the 2-dimensional reference spectra for $e^{\pm}$ and the background are fitted to data in the (TRD estimator–log(E/p)) plane by varying the normalizations of the signal and the background. The fit is performed for positive and negative rigidity data samples yielding, respectively, the numbers of positrons and electrons.

Several sources of systematic uncertainty have been estimated. At low energy, the most important contribution comes from the asymmetric acceptance of e^+ and e^-; at the highest energies, the uncertainty on the charge confusion between electrons and positrons is the dominant effect. The uncertainty on the reference spectra, the influence of the selection cuts and bin-to-bin migration in the energy spectrum are found to be subdominant at all energies. Above $\sim$10 GeV the measurement of the positron fraction is limited by statistics.

The measured positron fraction is presented in Fig. 2 as a function of the reconstructed energy at the top of the AMS detector. The result is compared with the most recent measurements by PAMELA [15] and Fermi-LAT [16]. The observation that the positron fraction is steadily increasing from 10 to 250 GeV is not consistent with only the secondary production of positrons [17].

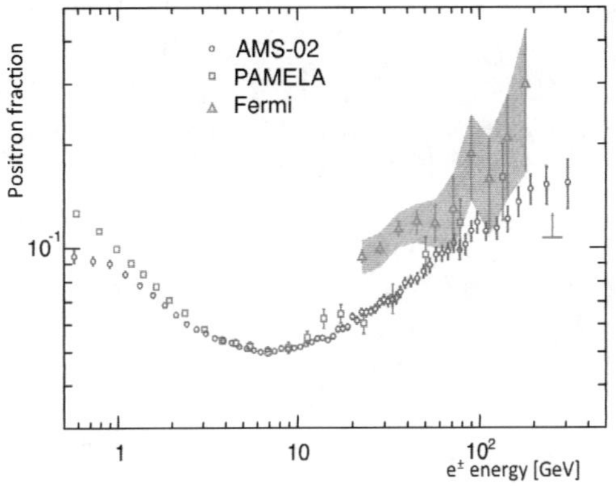

Figure 2. The positron fraction compared with the most recent measurements from PAMELA [15] and Fermi-LAT [16].

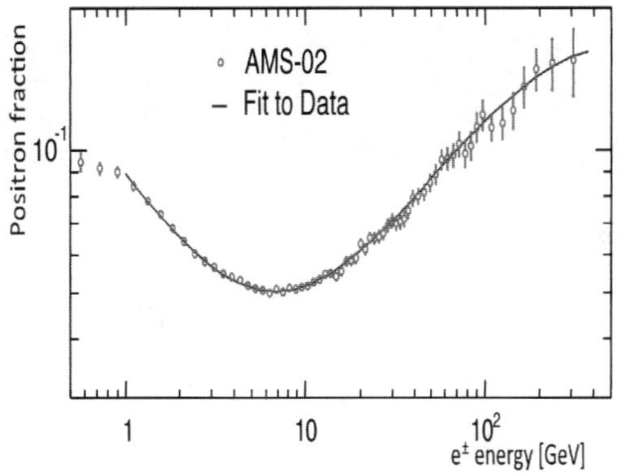

Figure 3. The positron fraction measured by AMS fit with the minimal model.

It is convenient to present the results comparing our data with a minimal model, where the e^+ and e^- fluxes, Φ_{e^+} and Φ_e, are parametrized as the sum of individual diffuse power law spectra and the contribution of a single common source of $e^\pm$:

$$\Phi_{e^+} = C_{e^+} E^{-\gamma_{e^+}} + C_s E^{-\gamma_s} e^{-E/E_s}$$

$$\Phi_{e^-} = C_{e^-} E^{-\gamma_{e^-}} + C_s E^{-\gamma_s} e^{-E/E_s}$$

where $C_{e^\pm}$ and C_s are the relative contributions from the diffuse and single source spectra. The diffuse spectra are assumed to follow power laws with spectral indices $\gamma_{e^\pm}$ whereas the source spectrum is described as a single power law with spectral index γ_s and an exponential cutoff with a characteristic energy E_s. With this parametrization the positron fraction depends on 5 parameters. The result of a fit to the data in the energy range 1 to 350 GeV based on the number of events in each bin is shown in Fig. 3 and yields a χ^2 of 28.7 for 57 degrees of freedom [1], with a diffuse positron spectrum softer than the diffuse electron spectrum ($\gamma_{e^-} - \gamma_{e^+} = -0.63 \pm 0.03$), a source spectrum harder than the diffuse electron spectrum ($\gamma_{e^-} - \gamma_s = 0.66 \pm 0.05$), a weight for

[1] For statistical and systematic errors added in quadrature. A χ^2 value of 37.9 is obtained if only statistical errors are considered.

the diffuse positron flux about ~10% of that of the diffuse electron flux ($C_{e^+}/C_{e^-} = 0.091 \pm 0.001$) and a weight for the common source amounting ~1% of that of the diffuse electron flux ($C_s/C_{e^-} = 0.0078 \pm 0.0012$). Finally, the fit yields $1/E_s = 0.0013 \pm 0.0007 \text{GeV}^{-1}$, corresponding to a cutoff energy of $760^{+1,000}_{-280}$ GeV.

The excellent agreement between the data and the model shows that the positron fraction spectrum is consistent with $e^{\pm}$ fluxes each of which is the sum of its diffuse spectrum and a single common power law source. No fine structures are observed in the data. This study also shows that the slope of the positron fraction as a function of energy decreases by an order of magnitude from 20 to 250GeV.

4. Search for positron anisotropy

The measurement of the positron fraction shows that the positron fraction is not consistent with the sole secondary production of positrons, but requires the inclusion of primary sources, whether from a particle physics or an astrophysical origin. Primary sources of cosmic ray positrons and electrons may induce some degree of anisotropy on the measured positron and electron fluxes.

The study of anisotropies in the positron fraction requires the selection of pure electron and positron samples. This is achieved with the additional requirement of a good energy-momentum matching and explicit cuts on the ECAL and TRD estimators. The overall selection efficiency for positrons and electrons is estimated to be above 80% in the acceptance of the ECAL. The remaining sample contains ~35,000 primary positrons, ~460,000 electrons with energies above 16 GeV and a negligible amount of protons.

The arrival directions of electrons and positrons are used to build sky maps in galactic coordinates, (b, l), containing the number of observed positrons and electrons. The maps corresponding to electrons and positrons in the energy range from 16 to 350 GeV are displayed in Fig. 4. The spread of the number of events collected on different directions is a consequence of the non uniform sky coverage of the AMS exposure.

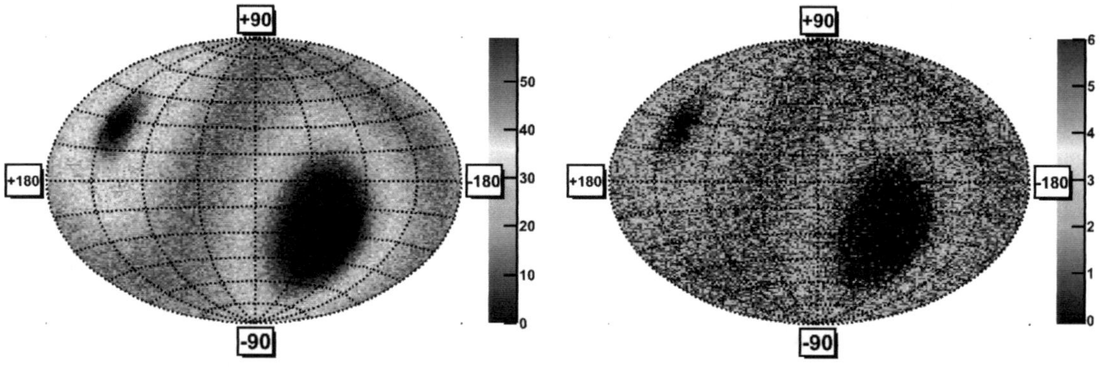

Figure 4. Sky maps showing the arrival directions of selected 16–350 GeV electrons (left) and positrons (right) in galactic coordinates using a Hammer-Aitoff projection. The color code reflects the number of events per bin.

For a given energy range, the positron to electron ratio is computed on each galactic coordinate bin. The consistency of the set of bin-to-bin ratios to a common value is estimated using a χ^2 test. A good agreement is found for all energy ranges. Moreover, no structure is observed on the projections along galactic latitude or longitude.

A general description of the relative fluctuations on the observed positron ratio is obtained by means of a spherical harmonic expansion:

$$\frac{r_e(b,l) - \langle r_e \rangle}{\langle r_e \rangle} = \sum_{\ell=0}^{\infty} \sum_{m=-\ell}^{\ell} a_{\ell m} \, Y_{\ell m}(\pi/2 - b, l),$$

where $r_e(b,l)$ denotes the positron ratio at (b,l), $\langle r_e \rangle$ is the average ratio over the sky map, $Y_{\ell m}$ are the real spherical harmonic functions, and a_{lm} are their corresponding amplitudes.

The amplitudes of spherical harmonic contributions at fixed angular scale, ℓ, are fit to the data independently for dipole ($\ell = 1$), quadrupole ($\ell = 2$) and octopole ($\ell = 3$) contributions with a χ^2 minimization. No significant a_{lm} is found at any angular scale ($\ell = 1, 2, 3$) and energy range. As an example, in Fig. 5 the results corresponding to a dipole contribution perpendicular to the galactic plane, a_{10}, are displayed together with the 1, 2 and 3σ contours as a function of the minimum energy. Similar sensitivity is obtained on the amplitudes of the other spherical harmonic contributions.

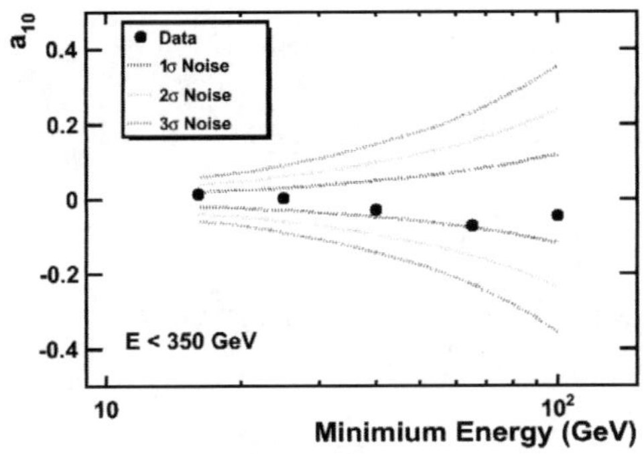

Figure 5. Amplitudes a_{10} obtained from fits of a dipole contribution to the data on different energy ranges. The dashed lines correspond to the 1, 2 and 3σ contours.

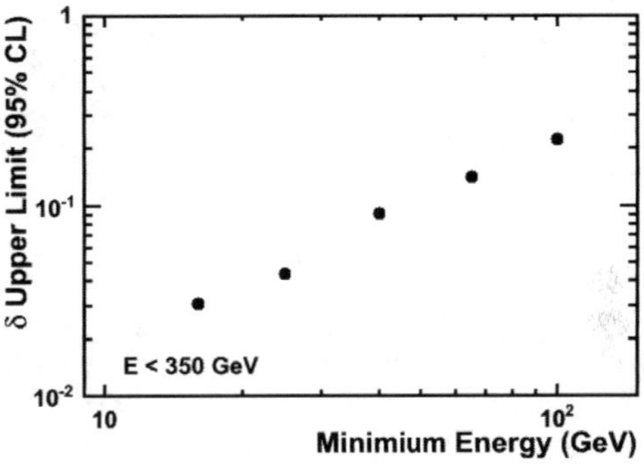

Figure 6. AMS upper limits on the dipole anisotropy parameter δ at the 95% confidence level on different energy ranges.

The intensity of the fluctuations on the studied angular scales is quantified with the coefficients of the angular power spectrum defined as $C_\ell = \frac{1}{2\ell+1} \sum_{m=-\ell}^{\ell} a_{\ell m}^2$. The values of the coefficients C_1, C_2 and C_3 obtained from the fits are consistent with the expectations from pure statistical fluctuations on all energy ranges.

The anisotropy induced by primary sources is expected to follow a dipole pattern with the maximum pointing towards the source and the minimum to the opposite direction. It is then customary to define the dipole anisotropy parameter, δ, as the relative difference between maximum and minimum amplitudes. Therefore, δ can be derived from the coefficient C_1 with the expression $\delta = 3\sqrt{C_1/4\pi}$.

Since, for all energy ranges, the coefficients of the multipole expansion are consistent with the expectations from isotropy, limits on the coefficients of the angular power spectrum, C_l, can be obtained. In particular, limits on the dipole anisotropy parameter δ are derived for any axis in galactic coordinates. The upper limits at the 95% confidence level for the 5 cumulative energy ranges are shown in Fig 6. The limit obtained for the energy range from 16 to 350 GeV is $\delta < 0.030$.

The stability of the results is verified by repeating the analysis on sky maps constructed with different binning and by changing the requirement on the minimum number of electrons per bin in a wide range. In all cases, negligible differences are found.

Journal of Physics: Conference Series **531** (2014) 012007 doi:10.1088/1742-6596/531/1/012007

The effect of a possible anisotropy in the electron flux is investigated by repeating the analysis on the positron to proton ratio. In addition, the influence of geomagnetic effects can be estimated on this ratio with the use of the directions obtained after backtracing in the geomagnetic field to the border of the Earth's magnetosphere [2] [18]. The sensitivity to a dipole anisotropy using the positron to proton ratio is found to be compatible with that obtained on the positron to electron analysis and similar sensitivity to a dipole anisotropy is found after backtracing.

Dark matter annihilation or decay in the Sun vicinity [12] could generate a relative positron excess towards the Sun direction. This effect is enhanced on the appropriate reference system. A similar analysis is performed using sky maps built in geocentric solar ecliptic coordinates, where the Earth-Sun axis defines its primary direction and its equatorial plane lies on the ecliptic. No significant deviation from isotropy is found and similar limits on a dipole anisotropy as those obtained in the analysis on galactic coordinates are derived. A complementary approach based on the methods for extended source evaluation in γ-ray astronomy [19] has been applied to the search for a local excess towards the Sun. No significant excess is found.

5. Discussion

Within the framework of the minimal model described in Sec. 3, AMS measurements of the positron fraction and the $e^+ + e^-$ flux [6] provide a complete description of the source component. The predictions of the different scenarios accounting for the positron excess can thus be directly validated.

For each particular scenario, the predictions for the positron source component and its distribution across the Galaxy can be computed. The propagation of the cosmic positrons from their sources to the heliosphere requires solving the transport equation which accounts for the diffusion, convection and energy losses. An analytical solution of the propagation equation for the primary sources can be obtained using the Green functions formalism as described in [21], where the two main processes, diffusion and energy losses, are considered.

Two different potential sources of high energy positrons have been investigated: Dark matter annihilation in the halo of the Galaxy and positron production in standard astrophysical objects.

Dark matter (DM) annihilation or decay can produce a large amount of positrons and electrons, hence, it constitutes a very attractive candidate to explain the reported excesses. However, conventional WIMP annihilation rates with canonical thermally averaged cross sections $\langle\sigma v\rangle = 3 \times 10^{-26}\mathrm{cm}^3/\mathrm{s}$ are too small to account for the observations. Moreover, the absence of an excess in antiprotons indicates that the annihilation through hadronic channels must be strongly suppressed.

A scan in the $M_{DM} - \langle\sigma v\rangle$ parameter space for a model-independent dark matter scenario, where DM particles annihilate into $\mu\bar{\mu}$ or $\tau\bar{\tau}$ final states, has been carried out. The mass distribution in the Galaxy halo is taken from the results obtained on recent N-body simulations [22, 23]. The resulting fluxes have been used to perform joint fits to the AMS positron fraction and $e^+ + e^-$ flux data for energies above 30 GeV, where solar modulation can be safely neglected. AMS data can be reproduced with thermally averaged annihilation cross sections $\sim 10^{-23}\mathrm{cm}^3/\mathrm{s}$ for $\sim$TeV DM mass candidates [20].

Standard astrophysical sources, such as SNR and pulsars, can also contribute to the observed positron excess. A pulsar model has been used to illustrate this scenario. The contribution from all gamma ray pulsars in the ATNF pulsar catalog [3] to the positron flux has been fit to the AMS data. The contribution from individual sources is gauged by their conversion efficiency, $\eta_{e^\pm}$, of the total spin-down power into $e^\pm$. The pulsars that contribute most to the positron flux are Geminga and Monogem, due to their distance and age. For them, best fits are obtained with

[2] http://www.geomagsphere.org/
[3] http://www.atnf.csiro.au/research/pulsar/psrcat

Journal of Physics: Conference Series **531** (2014) 012007 doi:10.1088/1742-6596/531/1/012007

$\eta_{e^\pm} \sim 16\%$ and $\eta_{e^\pm} \sim 4\%$ respectively with a negligible contribution from other pulsars [20].

The positron sources from these two scenarios are compared to the minimal model fit to AMS data in Fig. 7. A good agreement is found for both the pulsar and DM origin in the energy range of the AMS measurement where the minimal model uncertainty is small. Nevertheless small differences are apparent in regions where the minimal model presents large uncertainties. At low energies, a systematic undershoot in the physics models is observed, that could be an indication of either an incomplete description of the diffusion setup, solar modulation effects, or could even point out the presence of additional sources such as distant objects contributing to the low energy range. At very high energies near the source cutoff energy, the DM and pulsar models underestimate the source contribution to the total flux, which could reflect the presence of unaccounted young pulsars in a pulsar scenario, or a larger DM mass in a dark matter scenario.

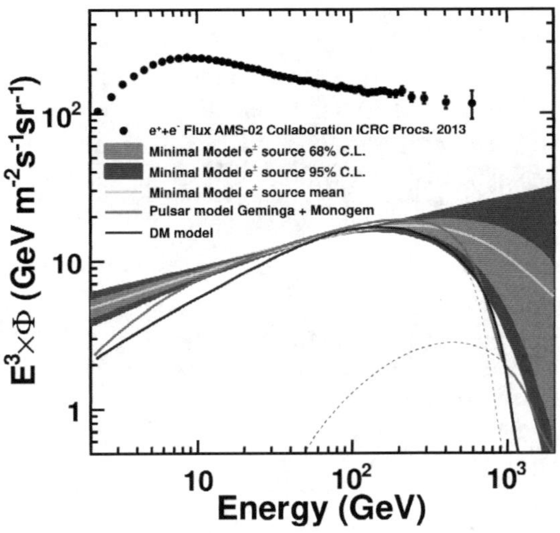

Figure 7. AMS e$^+$+e$^-$ flux measurement [6], together with the positron source contribution within the minimal model described in the text (green line and blue shaded areas). Best fits for a DM model (blue line) and a pulsar model (red line) with Geminga (red dashed line) and Monogem (blue dashed line) contributions are also displayed.

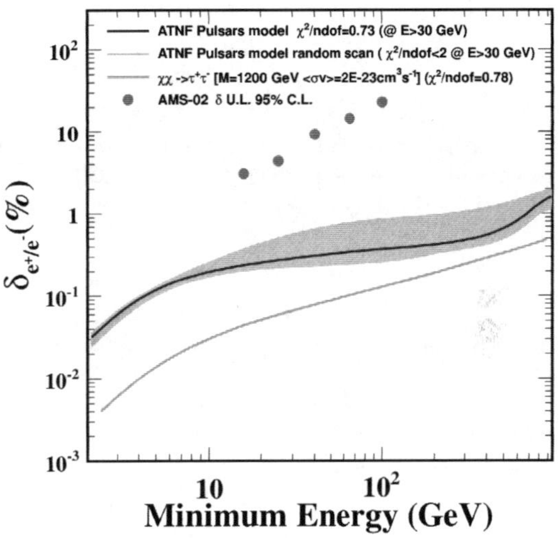

Figure 8. AMS upper limits on the dipole anisotropy parameter δ at the 95% confidence level on different energy ranges. The predictions from the best fit DM (brown line) and pulsar models (blue line and gray shaded area) are also shown.

Given that for both production mechanisms the sources are distributed in the Galaxy with some spatial structure, small anisotropies should be present in the arrival directions. A residual dipole anisotropy in the source positron flux for both pulsar and dark matter models can be computed after their diffusion in the Galaxy [24], and the degree of anisotropy in the positron to electron ratio can be estimated taking into account the isotropic background for both positrons and electrons.

The expected dipole anisotropy for the different production mechanisms is shown in Fig. 8 together with the upper limits derived int Sec. 4. Experimental limits are well above theoretical expectations and therefore both models are compatible with the observations [20].

6. Summary and outlook

The precise measurements of cosmic ray positrons and electrons performed by AMS on the ISS indicate a steady increase in the positron fraction from 10 to 250 GeV with no spectral structure and no observable anisotropy. This observation is not consistent with the sole secondary production of positrons and requires the inclusion of primary sources, whether from a particle physics or an astrophysical origin.

Models reflecting these two possible scenarios have been tuned to reproduce the positron source flux derived from the AMS positron fraction and the $e^+ + e^-$ flux measurements. A DM inspired model with enhanced leptonic decays is able to reproduce the positron source term at the expense of a boost in the thermally averaged cross section. On the other hand, a pulsar model which incorporates the contribution from most relevant individual sources is also compatible with the AMS data, provided that their conversion efficiency of spin-down power into $e^\pm$ is $\sim$10%. Both scenarios predict a degree of dipole anisotropy on the positron to electron ratio compatible with the AMS limits.

In spite of its unprecedented precision, the uncertainty on the AMS measurement of the positron fraction and its anisotropy is dominated by the limited statistics. Current results have been obtained on less that 10% of the expected sample as AMS will continue collecting data for more than 10 years. This will allow clarifying the high energy behavior while extending the energy range towards 1 TeV as well as improving its sensitivity to a dipole anisotropy to 1%.

References

[1] A. Kounine, Int. J. Mod. Phys. **E 21** (2012) 123005.
[2] M. Aguilar *et al.*, Phys. Rev. Lett. **A 110** (2013) 141102.
[3] A. Kounine, for the AMS Collab., Proc. 33rd ICRC, Rio de Janeiro, Brazil (2013).
[4] J. Casaus, for the AMS Collab., Proc. 33rd ICRC, Rio de Janeiro, Brazil (2013).
[5] S. Schael, for the AMS Collab., Proc. 33rd ICRC, Rio de Janeiro, Brazil (2013).
[6] B. Bertucci, for the AMS Collab., Proc. 33rd ICRC, Rio de Janeiro, Brazil (2013).
[7] S. Haino, for the AMS Collab., Proc. 33rd ICRC, Rio de Janeiro, Brazil (2013).
[8] V. Choutko, for the AMS Collab., Proc. 33rd ICRC, Rio de Janeiro, Brazil (2013).
[9] A. Oliva, for the AMS Collab., Proc. 33rd ICRC, Rio de Janeiro, Brazil (2013).
[10] A. K. Harding and R. Ramaty, Proc. 20th ICRC, Moscow, USSR (1987); D. Hooper, P. Blasi and P. Serpico, JCAP **01** (2009) 25; D. Grasso *et al.*, Astropart. Phys. **32** (2009) 140; S. Profumo, CEJP **10** (2011) 1-31.
[11] J. Silk and M. Srednicki, Phys. Rev. Lett. **53** (1984) 624; J. R. Ellis *et al.*, Phys. Lett. **B214** (1988) 403; M. Cirelli *et al.*, Nucl. Phys. **B813** (2009) 1.
[12] P. Schuster *et. al.*, Phys. Rev. **D82** (2010) 115012.
[13] I. Büsching *et at.*, Astrophys. J. **678** (2008) L3; I. Cernuda, Astropart. Phys. **34** (2010) 59.
[14] B. Roe *et al.*, Nucl. Instrum. Meth. **A 543** (2005) 577.
[15] P. Picozza, Proc. 4th Intl. Conf. on Particle and Fundamental Physics in Space, Geneva, Switzerland (2012); O. Adriani *et al.*, Astropart. Phys. **34** (2010) 1; O. Adriani *et al.*, Nature **458** (2009) 607.
[16] M. Ackermann *et al.*, Phys. Rev. Lett. **108** (2012) 011103.
[17] P.D. Serpico, Astropart. Phys. **39-40** (2012) 2; T. Delahaye *et al.*, Astron. Astrophys. **501** (2009) 821; I. Moskalenko and A. Strong, Astrophys. J. **493** (1998) 693.
[18] P. Bobik et al., Proc. 33rd ICRC, Rio de Janeiro, Brazil (2013); M. Boschini et al., *ibid.*
[19] T. Li and Y. Ma, Astrophys. J. **272** (1983) 317.
[20] I. Cernuda, *Cosmic-Ray Positron Fraction Measurement with the AMS-02 Detector.* PhD Thesis. Universidad Complutense de Madrid, Spain (2014).
[21] T. Delahaye *et al.*, Phys. Rev. **D77** (2008) 063527.
[22] V. Springel *et at.*, Month. Not. Roy. Astron. Soc. **391** (2008) 1685.
[23] L. Pieri *et al.*, Phys. Rev. **D83** (2011) 023518.
[24] C.Y. Mao and C.S Shen, Chinese Journal of Physics **10** (1972) 16.

Cosmic Ray Anisotropy Workshop 2013 (CRA2013) IOP Publishing
Journal of Physics: Conference Series **531** (2014) 012008 doi:10.1088/1742-6596/531/1/012008

A hadronic explanation of the lepton anomaly

Philipp Mertsch
Kavli Institute for Particle Astrophysics & Cosmology, 2575 Sand Hill Road, M/S 29, Menlo Park, CA 94025, USA
E-mail: pmertsch@stanford.edu

Subir Sarkar
Rudolf Peierls Centre for Theoretical Physics, University of Oxford, 1 Keble Road, Oxford OX1 3NP, UK
Niels Bohr Institute, Copenhagen University, Blegdamsvej 17, 2100 Copenhagen Ø, Denmark

Abstract. The rise in the positron fraction, observed by PAMELA, *Fermi*-LAT and most recently by AMS-02, has created a lot of interest, fuelled by speculations about an origin in dark matter annihilation in the Galactic halo. However, other channels, e.g. antiprotons or gamma-rays, now severely constrain dark matter interpretations, thus requiring astrophysical sources of positrons. We have investigated the possibility that supernova remnants, the most likely sources of Galactic cosmic rays, can in fact also produce a hard spectrum of secondary positrons, by spallation and acceleration at the shock. This mechanism is guaranteed if hadronic CRs are present and would also lead to observable signatures in other secondary channels like the boron-to-carbon or antiproton-to-proton ratios. If such features were borne out by upcoming AMS-02 data, this would rule out other explanations.

1. Introduction

In propagation models of galactic cosmic rays (GCRs), the positron fraction, i.e. the ratio of positrons to the sum of electrons and positrons, is expected to fall at all energies. Its rise, first observed by PAMELA [1], later confirmed by *Fermi*-LAT [2] and most recently by AMS-02 [3], has been widely speculated to be of exotic origin, namely due to annihilation [4] or decay [5] of dark matter (DM) in the Galactic halo. However, since the discovery of this positron anomaly, a lot of effort has been devoted to testing this possibility in other channels with the result that DM interpretations are now severely constrained: A related rise in the antiproton-to-proton ratio ($\bar{\text{p}}/\text{p}$) has not been observed, thus constraining the coupling of DM to quarks [6]. The non-observation of inverse Compton gamma-rays at intermediate latitudes from additional CR electrons and positrons severely constrains the annihilation of DM into leptons [7]. Finally, the annihilation of DM into lepton final states can also be tested in the CMB which is now setting competitive bounds on the annihilation cross-section in the relevant mass range (see, e.g. [8]).

Among the astrophysical interpretations—see [9] for an overview—pulsar wind nebulae (PWNe) have received a lot of attention. Electrons are extracted from the surface of the neutron star in intense magnetic fields, get accelerated in the magnetosphere and their curvature radiation creates electron-positron pairs which can be further accelerated by the nebula's termination shock. Whereas the spin-down power and age (and possibly also the distance)

Journal of Physics: Conference Series **531** (2014) 012008 doi:10.1088/1742-6596/531/1/012008

can be determined from radio observations, the efficiency of electron-positron production, the spectral index and the cut-off energy are mostly free parameters and are thus oftentimes fixed by hand to fit the observed lepton fluxes. Modelling of the lepton data has been attempted under the alternative assumptions that either only one (at most a few) nearby pulsar(s) or the whole galactic population of pulsars contribute(s). It has been argued [10, 11, 12] that a clear indication of this model would be the observation of a dipole anisotropy (in positrons) in the direction of the nearest pulsar. There is, however, a caveat. For a fixed spatial distribution of the CR density, the amplitude of the anisotropy is proportional to the *local* diffusion coefficient whereas what is obtained from measurements of secondary-to-primary ratios, like boron-to-carbon (B/C), is the diffusion coefficient, spatially averaged over large regions in the Galaxy. Furthermore, in the last couple of years it has become apparent that anisotropies in hadronic CRs are not well understood: The dipole anisotropy observed between tens of TeV and up to a few PeV is much smaller than expected from diffusion models. Even more puzzling, the sky maps of arrival directions seem to contain angular power on much smaller scales, possibly pointing at the limitations of diffusion models. We conclude that the observation of a dipole anisotropy in leptons, related to the presence of a nearby PWN is very uncertain and therefore cannot be used as a test or even diagnostic of such a model.

In the following, we focus on a particular minimal and testable model [13, 14]: minimal in so far as no new class of CR sources is required; testable because the rise in the positron fraction is mirrored in similar features in other secondary-to-primary ratios, like B/C [15, 16, 17] or $\bar{\mathrm{p}}/\mathrm{p}$ [18, 17]. It is commonly assumed that supernova remnants (SNRs), the most likely candidate sources are depleted of certain species, so-called secondaries, which only get produced by spallation during the propagation of primaries through the interstellar medium (ISM). The amount of secondaries produced *inside* the sources is usually ignored since the grammage experienced by primaries inside the source is much smaller than the grammage experienced in the Galaxy (see, however, [19]). What this argument ignores is the possibility that the secondaries could have a harder spectrum than their parent primaries which could lead to an observable secondary spectrum at higher energies despite the small total number. Here, we revisit this model in light of new AMS-02 data [20, 21, 22, 23]. To avoid possible inconsistencies between results from different experiments, we focus on AMS-02 data alone.

2. Acceleration of secondaries

In the following we briefly recap the results for the spectrum of primaries and secondaries at a SNR shock in the test-particle approximation; for details see, e.g. [17]. The usual diffusion-convection equation is separately solved in the upstream ($x < 0$) and downstream ($x > 0$) of a SNR shock with shock speed u_-. In the test-particle approximation, the ratios of densities and speeds upstream and downstream, i.e. the compression factor, $r = n_+/n_- = u_-/u_+$, determines the spectral index of primaries, $\gamma = 3r/(r-1)$. The steady-state phase-space density at the shock computes as

$$f_i^0(p) = \int_0^p \frac{\mathrm{d}p'}{p'} \left(\frac{p'}{p}\right)^\gamma \mathrm{e}^{-\gamma(1+r^2)(D_i(p)-D_i(p'))\Gamma_i^-(p)/u_-^2} \left[\gamma(1+r^2)\frac{D_i(p')}{u_-^2} q_i^0(p') + \gamma Y_i \delta(p'-p_0)\right] , \quad (1)$$

and the spatial dependence on the downstream side is given by

$$f_i^+(x,p) = f_i^0(p) + r(q_i^0(p) - \Gamma_i^- f_i^0(p))\frac{x}{u_+} . \quad (2)$$

Here, D_i is the diffusion coefficient, Γ_i^- the rate of inelastic collisions upstream and $q_i^0 = \sum_{j>i} c\beta_j n_- \sigma_{j\to i} f_j^0$ is the source term for secondary species i by spallation of heavier species

Table 1. Parameter values of the two models adopted in our analysis, both for the source (K_B) and for the galactic propagation (δ, κ, $(\mathrm{d}v/\mathrm{d}z)$ and $z_{\max}$).

	K_B	δ	κ $[10^{28}\,\mathrm{cm}^2/\mathrm{s}]$	$(\mathrm{d}v/\mathrm{d}z)$ $[\mathrm{km/s\,kpc}]$	$z_{\max}$ [kpc]
Model 1	2,3,4	0.75	2	10	3
Model 2	9	0.65	0.6	8	1

j, with n_- the upstream gas density and $\sigma_{j\to i}$ the spallation cross section; Y_i is the abundance of primary i far upstream. It can be seen from eq. 1 that the spectrum of secondaries at the shock is harder than the injection spectrum (or the primary spectrum) due to the diffusion coefficient.

We make the simplifying assumption that after the lifetime τ_{SNR}, the integrated downstream spectrum,

$$\frac{\mathrm{d}N_i}{\mathrm{d}p} = 4\pi \int_0^{\tau_{\mathrm{SNR}} u_+} \mathrm{d}x\, x^2 4\pi p^2 f_i(x,p) = 4\pi p^2 V \left[f_i^0 + \frac{3}{4} r\, \tau_{\mathrm{SNR}} \left(q_i^0 - \Gamma_i^- f_i^0 \right) \right] , \qquad (3)$$

is released into the ISM. We have modelled the transport in the ISM using the `GALPROP` code [24] which numerically solves the transport equation for all CR species, including the production of secondaries due to spallation on the interstellar gas. As the production of secondaries in SNRs affects the secondary-to-primary ratios, like B/C, that are usually used to determine the propagation parameters, i.e. spatial diffusion coefficient κ (at a reference rigidity 4 GV), its spectral index δ or the gradient of the Galactic wind, $(\mathrm{d}v/\mathrm{d}z)$, we cannot rely on earlier determinations of these parameters. Instead we adopt a set of parameters that allows us to reproduce all AMS-02 measurements: The half-height of the diffusion volume, $z_{\max}$, is fixed to 3 kpc for model 1, and to 1 kpc for model 2. Solar modulation is treated in the simplified force-field approximation, allowing for different modulation potentials for different species. See Tbl. 1 for the other propagation parameters.

Among the free parameters of the source spectrum, i.e. r, $u_\pm$, τ_{SNR}, n_- as well as the diffusion coefficient $D = \beta c r_L(p)/3 \simeq 3\times 10^{22}\, K_B\,(pc/\mathrm{GeV})\, Z^{-1} B_{\mu\mathrm{G}}^{-1}\,\mathrm{cm}^2\,\mathrm{s}^{-1}$, where $r_{\mathrm{L}}(p)$ is the Larmor radius, only the combination $K_B/(u_-^2 B)$ enters into the secondary terms. We therefore fix some of the parameters, i.e. $B_{\mu\mathrm{G}} = 1$ and $u_- = 5\times 10^7\,\mathrm{cm\,s}^{-1}$, $n_- = 2\,\mathrm{cm}^{-3}$ while allowing K_B and τ_{SNR} to vary. For both models, we find $\tau_{\mathrm{SNR}} = 4\times 10^4\,\mathrm{yr}$ to give a good fit; K_B is shown with the other parameters in Tbl. 1. To reproduce the absolute proton and helium fluxes, the spectral index γ is also allowed to vary, but we find that $r \approx 4$ is sufficient. We note that a slightly harder Helium injection is required by the AMS-02 data which was already apparent in the PAMELA data [25]. For the primary electron injection spectrum a break has been inferred from radio data [26]; here we adopt a power-law slope of 1.6 below and 2.55 (2.65) above the break for model 1 (2). We furthermore fix the maximum energy of accelerated leptons to $E_{\max} = 10\,\mathrm{TeV}$.

Above $\sim 100\,\mathrm{GeV}$, the diffusion-loss length for leptons becomes shorter than the average distance between sources, such that lepton fluxes are very much dependent on the exact configuration of sources which is *a priori* unknown. However, one can compute statistical quantities like the expectation value from an ensemble of source distributions and the fluctuations around it. To this end we have employed the 3-dimensional, stochastic SNR mode of `GALPROP`, generating the fluxes for 25 realisations of a pulsar-like [27] source distribution. The shaded regions in the following figures represent the envelope of these fluxes.

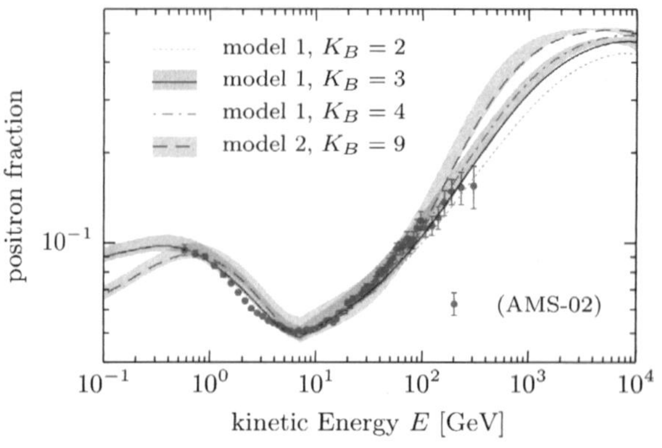

Figure 1. The positron fraction, $e^+/(e^+ + e^-)$ measured by AMS-02 [3] (circles) and for the acceleration of secondaries model with different values of the diffusion rate near the SNR shock.

3. Results

Fig. 1 shows the main result of the model, the positron fraction. It can be seen that in both models, it is falling at low energies, has a minimum around 7 GeV and rises at higher energies, in agreement with the latest AMS-02 data. Compared to earlier works [13, 18, 15, 14], the relatively milder rise in the positron fraction allows to consistently fit the absolute electron and positron fluxes, see Fig. 2. We emphasise that once the propagation parameters have been fixed,

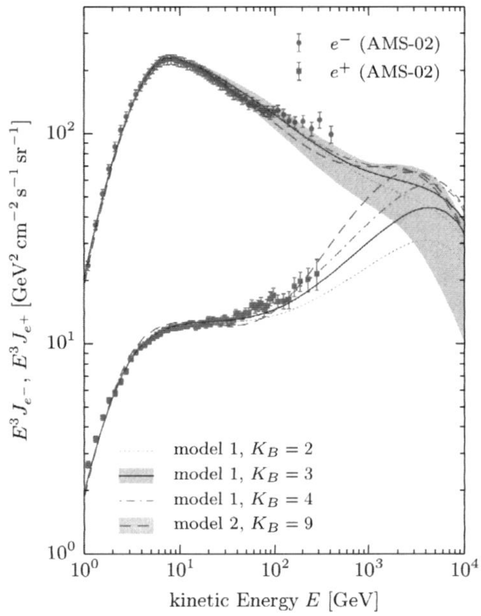

Figure 2. Absolute electron and positron fluxes measured by AMS-02 [20] (circles and squares, respectively) and for the acceleration of secondaries model.

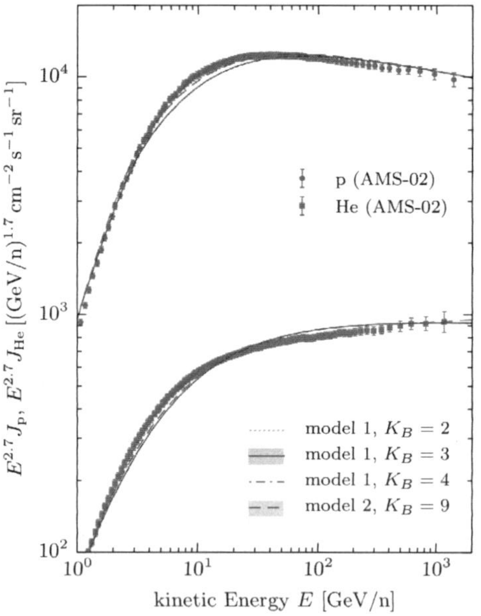

Figure 3. Absolute proton and helium fluxes measured by AMS-02 [21, 22] (circles and squares, respectively) and for the acceleration of secondaries model.

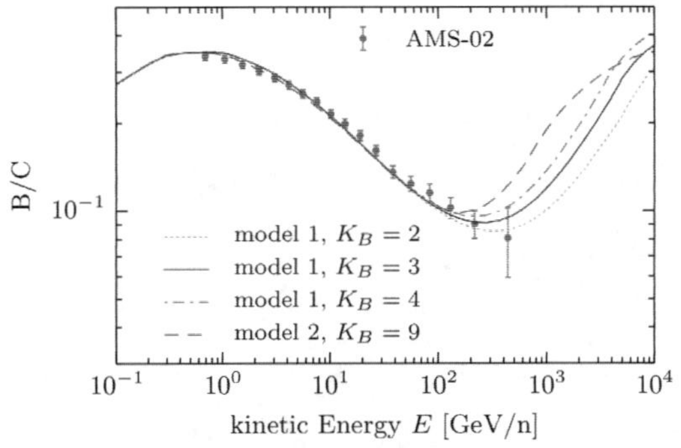

Figure 4. The boron-to-carbon ratio measured by AMS-02 [23] (circles) and for the acceleration of secondaries model.

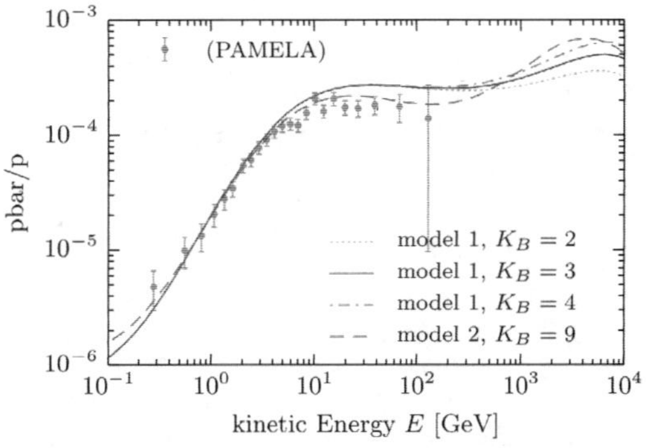

Figure 5. The antiproton-to-proton ratio measured by PAMELA [28] (circles) and for the acceleration of secondaries model.

the absolute normalisation of the electron and positron fluxes is constrained by the proton and Helium fluxes, see Fig. 3. Requiring to reproduce both at the same time therefore constitutes a non-trivial cross-check. In Figs. 4 and 5 we show the model prediction for B/C and $\bar{\mathrm{p}}/\mathrm{p}$ which are affected by the production and acceleration of secondaries in a similar fashion as the secondary positrons and electrons. Two differences are however noteworthy: First, whereas the positron fraction has a minimum around 7 GeV, B/C only starts rising at hundreds of GeV. This is due to the different inelasticity of positron and boron production (positrons are on average produced at 1/20 of the proton energy whereas boron retains the energy per nucleon of the parent primary) and the break in the electron spectrum. Second, $\bar{\mathrm{p}}/\mathrm{p}$ exhibits a distinctive plateau between 10 GeV and hundreds of GeV. Whereas the contribution from the spallation in the ISM is falling at lower energies for the positron fraction or B/C, it is still rising below ~ 10 GeV for $\bar{\mathrm{p}}/\mathrm{p}$. The additional hard component from the secondaries produced and accelerated near the SNR shock therefore leads to a plateau rather than a minimum.

We note that unlike Ref. [29] we explored a larger parameter space for the propagation model, allowing for $\delta = 0.65 - 0.75$ as is expected in diffusion-convection models [30]. This allows us to fit the positron fraction without violating the constraints set by B/C or $\bar{\mathrm{p}}/\mathrm{p}$. The upcoming data on $\bar{\mathrm{p}}/\mathrm{p}$ from AMS-02 and future B/C data will allow to further test this minimal model and all other possible explanations of the rising positron fraction.

Acknowledgements

We are grateful to Stefan Schael for helpful discussions. PM is supported by DoE contract DE-AC02-76SF00515 and a KIPAC Kavli Fellowship. SS acknowledges a Niels Bohr Professorship and STFC UK for support.

References

[1] Adriani O *et al.* (PAMELA Collaboration) 2009 *Nature* **458** 607–609 (*Preprint* 0810.4995)
[2] Ackermann M *et al.* (Fermi LAT Collaboration) 2012 *Phys.Rev.Lett.* **108** 011103 (*Preprint* 1109.0521)
[3] Aguilar M *et al.* (AMS Collaboration) 2013 *Phys.Rev.Lett.* **110** 141102
[4] Bergstrom L, Bringmann T and Edsjo J 2008 *Phys.Rev.* **D78** 103520 (*Preprint* 0808.3725)
[5] Nardi E, Sannino F and Strumia A 2009 *JCAP* **0901** 043 (*Preprint* 0811.4153)
[6] Evoli C, Cholis I, Grasso D, Maccione L and Ullio P 2012 *Phys.Rev.* **D85** 123511 (*Preprint* 1108.0664)
[7] Ackermann M *et al.* (LAT collaboration) 2012 *Astrophys.J.* **761** 91 (*Preprint* 1205.6474)
[8] Madhavacheril M S, Sehgal N and Slatyer T R 2013 (*Preprint* 1310.3815)
[9] Serpico P D 2012 *Astropart.Phys.* **39-40** 2–11 (*Preprint* 1108.4827)
[10] Buesching I, de Jager O, Potgieter M and Venter C 2008 (*Preprint* 0804.0220)

[11] Profumo S 2011 *Central Eur.J.Phys.* **10** 1–31 (*Preprint* 0812.4457)
[12] Hooper D, Blasi P and Serpico P D 2009 *JCAP* **0901** 025 (*Preprint* 0810.1527)
[13] Blasi P 2009 *Phys.Rev.Lett.* **103** 051104 (*Preprint* 0903.2794)
[14] Ahlers M, Mertsch P and Sarkar S 2009 *Phys.Rev.* **D80** 123017 (*Preprint* 0909.4060)
[15] Mertsch P and Sarkar S 2009 *Phys.Rev.Lett.* **103** 081104 (*Preprint* 0905.3152)
[16] Tomassetti N and Donato F (collaboration) 2012 *Astron.Astrophys.* **544** A16 (*Preprint* 1203.6094)
[17] Mertsch P and Sarkar S 2014 (*Preprint* 1402.0855)
[18] Blasi P and Serpico P D 2009 *Phys.Rev.Lett.* **103** 081103 (*Preprint* 0904.0871)
[19] Berezhko E, Ksenofontov L T, Ptuskin V, Zirakashvili V and Voelk H 2003 *Astron.Astrophys.* **410** 189–198 (*Preprint* astro-ph/0308199)
[20] S Schael on behalf of the AMS Collaboration *Proc. 33rd Intern. Cosmic Ray Conf. (2013)*
[21] S Haino on behalf of the AMS Collaboration *Proc. 33rd Intern. Cosmic Ray Conf. (2013)*
[22] V Choutko on behalf of the AMS Collaboration *Proc. 33rd Intern. Cosmic Ray Conf. (2013)*
[23] A Oliva on behalf of the AMS Collaboration *Proc. 33rd Intern. Cosmic Ray Conf. (2013)*
[24] Moskalenko I and Strong A 1998 *Astrophys.J.* **493** 694–707 (*Preprint* astro-ph/9710124)
[25] Adriani O *et al.* (PAMELA Collaboration) 2011 *Science* **332** 69–72 (*Preprint* 1103.4055)
[26] Strong A, Orlando E and Jaffe T 2011 *Astron.Astrophys.* **534** A54 (*Preprint* 1108.4822)
[27] Lorimer D R 2003 (*Preprint* astro-ph/0308501)
[28] Adriani O *et al.* (PAMELA Collaboration) 2010 *Phys.Rev.Lett.* **105** 121101 (*Preprint* 1007.0821)
[29] Cholis I and Hooper D 2013 *Phys.Rev.D* (*Preprint* 1312.2952)
[30] Maurin D, Putze A and Derome L 2010 *Astron.Astrophys.* **516** A67 (*Preprint* 1001.0553)

Journal of Physics: Conference Series **531** (2014) 012009 doi:10.1088/1742-6596/531/1/012009

Large Scale Anisotropy of Cosmic Rays and Directional Neutrino Signals from Galactic Sources

Luis A. Anchordoqui[1], Haim Goldberg[2] Angela V. Olinto[3,4], Thomas C. Paul[1,2], Brian J. Vlcek[1] and Thomas J. Weiler[5]

[1] Department of Physics, University of Wisconsin-Milwaukee, Milwaukee, WI, 53201 USA
[2] Department of Physics, Northeastern University, Boston, MA 02115, USA
[3] Department of Astronomy and Astrophysics, Enrico Fermi Institute, Chicago, Il 60637, USA
[4] Kavli Institute for Cosmological Physics, University of Chicago, Chicago, Il 60637, USA
[5] Department of Physics & Astronomy, Vanderbilt University, Nashville, TN, 37235 USA

Speaker's E-mail: luis.anchordoqui@gmail.com

Abstract. We investigate the neutrino $\leftrightharpoons$ cosmic ray connection for sources in the Galaxy in terms of two observables: the shape of the energy spectrum and the distribution of arrival directions. We also study the associated gamma ray emission from these sources.

Above about 10 GeV, cosmic ray (CR) energy spectrum observed at Earth falls roughly as a power law, with flux decreasing almost three orders of magnitude per energy decade until eventually it exhibits a strong suppression near 60 EeV [1]. Closer examination reveals several other spectral features. A steepening of the spectrum between $J(E) \propto E^{-2.67\pm0.07}$ and $E^{-3.07\pm0.11}$ has been labeled the "knee" and is prominent at $E_{\rm knee} \approx 3$ PeV [2]. A less prominent "second knee", corresponding to additional softening of the spectrum, $J(E) \propto E^{-3.52\pm0.19}$, appears above 0.3 EeV [3]. At $E_{\rm ankle} \approx 3$ EeV a pronounced hardening of the spectrum appears, the so-called "ankle" feature [4].

The Earth lies near the edge of the Galaxy, so if CR sources are concentrated in the Galactic disk one would expect to observe a dipole anisotropy in the flux, with a relative excess in the direction of the Galactic center. Hints of this anisotropy have in fact emerged. The IceCube Collaboration has reported [5] an excess of events in the 400 TeV sky-map (29° smoothing) at right ascension $\equiv \alpha = 256.6°$ and declination $\equiv \delta = -25.9°$, with a pre-trial significance of 5.3σ. The data also show a deficit at $\alpha = 73.1°$ and $\delta = -25.3°$, with a pre-trial significance value of 8.6σ (optimized smoothing of 21°). After correcting for the trials, only the deficit remains significant beyond the 5σ level, with a post-trial significance of 6.3σ.

The right ascension distribution of the flux can be characterized by the amplitudes and phases of its Fourier expansion [6]

$$J(\alpha) = J_0[1 + \mathcal{A}\cos(\alpha - \phi) + \mathcal{A}'\cos(2(\alpha - \phi')) + \ldots], \qquad (1)$$

where J_0 is the monopole intensity, $\mathcal{A}$ and $\mathcal{A}'$ are the first and second harmonic amplitudes and ϕ and ϕ' the associated phases. The right ascension harmonic analysis cannot reveal the component of the anisotropy vector $\vec{\delta}$ along the Earth rotation axis, $\delta_{\|} = \vec{\delta}\sin\lambda$, where λ is the latitude of the direction where the flux is maximum. The first harmonic amplitude $\mathcal{A}$ is related to the component of anisotropy in the equatorial plane, $\delta_\perp = \vec{\delta}\cos\lambda$, via $\mathcal{A} \simeq \delta_\perp\ \overline{c_\delta}$,

Journal of Physics: Conference Series **531** (2014) 012009 doi:10.1088/1742-6596/531/1/012009

where $\overline{c_\delta}$ is the mean value of the cosine of the event declinations [7]. The first harmonic amplitude reported by the IceCube Collaboration is $\mathcal{A} = (3.7 \pm 0.7_{\rm stat.} \pm 0.7_{\rm stat}) \times 10^{-4}$ and the associated phase is $\phi = 239 \pm 10.6_{\rm stat.} \pm 10.8_{\rm syst.}$. These results are in agreement with previous observations by EAS-TOP Collaboration which, for 370 TeV, reported $\mathcal{A} = (6.4 \pm 2.5) \times 10^{-4}$ with $\phi = 204.0 \pm 22.5$ [8].

Recently, the Pierre Auger Collaboration reported an analysis of the first harmonic modulation in the right ascension distribution of the events recorded from 1/1/2004 to 12/31/2010 with the surface detector (SD) array, and from 9/12/2007 to 4/11/2011 with the infill array [9]. This analysis takes advantage of the wide range of energy (10 PeV – 100 EeV) that the Pierre Auger Observatory is able to scan thanks to the infill array. While no clear evidence for anisotropy has been found, yet it is interesting to note that in the range above 1 EeV, 3 out of the 4 energy bins are above the 99% C.L. expectation from isotropy, *i.e.* only one percent of isotropic samples would show equal or larger amplitudes. The phase evolution in this wide energy range has an interesting behavior, with a smooth transition from a common phase of $\phi = 270°$ in the bins below 1 EeV to a phase $\phi = 90°$ above 5 EeV. The phase at lower energies is compatible with the right ascension of the Galactic center, $\alpha_{\rm GC} \simeq 268.4°$. To test the hypothesis that the phase is undergoing a smooth transition, the Pierre Auger Collaboration began to independently analyze data obtained after April 2011. After 18 months the new and independent data set is showing a similar trend [9]. Another 18 months of data collection to reach an aperture of 21,000 $\rm km^2$ sr with the independent data set is needed before the trend can be confirmed. It is interesting to note that despite the possible hints for cosmic ray anisotropy discussed above, any such anisotropy would be remarkably small (at the % level). The existing limits on the equatorial amplitude $\delta_\perp$ and the reported phases by the various experiments are shown in Fig. 1.

To quantify the spectral features characteristic of Galactic CR models we adopt the "leaky box" picture, in which CRs propagate freely in the Galaxy, contained by the magnetic field but with some probability to escape which is constant in time. The local energy density is given by

$$n_{\rm CR}(E) \equiv \frac{4\pi}{c} J(E) \approx Q(E)\ \tau(E/Z), \tag{2}$$

where $Q(E) \propto E^{-\alpha}$ is the generation rate of primary CRs and $\tau(E/Z) \propto E^{-\delta}$ is the rigidity-dependent confinement time (for details, see e.g. [13]). Fits to the energy dependence of secondary to primary ratios yield $\delta = 0.6$ [14]. For a source index $\alpha \simeq 2.07$, which is close to the prediction of Fermi shock acceleration, inclusion of propagation effects reproduces the observed spectrum. However, $\delta = 0.6$ results in an excessively large anisotropy which is inconsistent with the upper limits shown in Fig. 1 [15]. Consistency with anisotropy can be achieved by adopting a Kolmogorov index, $\delta = 1/3$ [16, 17]. The apparent conflict with the secondary to primary composition analyses can be alleviated through small variations of the energy dependence of the spallation cross sections, or variation in the matter distribution in the Galaxy [16]. This hypothesis implies a steeper source spectrum, $\alpha \simeq 2.34$.

It is helpful to envision the CR engines as machines where protons are accelerated and (possibly) permanently confined by the magnetic fields of the acceleration region. The production of neutrons and pions and subsequent decay produces neutrinos, γ-rays, and CRs. If the CR source also produces high energy neutrinos, then pion production must be the principal agent for the high energy cutoff on the proton spectrum. Conversely, since the protons must undergo sufficient acceleration, inelastic pion production needs to be small below the cutoff energy; consequently, the plasma must be optically thin. Since the interaction time for protons is greatly increased over that of neutrons due to magnetic confinement, the neutrons escape before interacting, and on decay give rise to the observed CR flux. The foregoing can be summarized as three conditions on the characteristic nucleon interaction time scale $\tau_{\rm int}$; the neutron decay

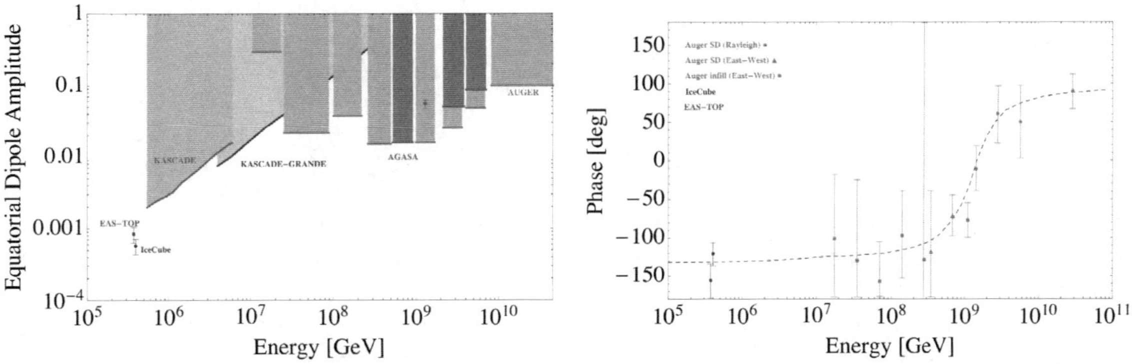

Figure 1. Left: Measurements of the first harmonic amplitude (corrected by the mean value of the cosine of the event declinations) by IceCube [5], EAS-TOP [8], and AGASA [10]. The shaded regions are excluded by null results of searches by KASCADE (@ 95% C.L.) [11] and KASCADE-Grande (@ 95% C.L.) [12], Auger (@ 99% C.L.) [9], and AGASA collaborations [10]. **Right:** Phase of the first harmonic as a function of energy as reported by the IceCube [5], the EAS-TOP [8], and the Pierre Auger [9] collaborations. The phase shows a smooth evolution from the Galactic center towards the Galactic anti-center region.

lifetime τ_n; the characteristic cycle time of confinement $\tau_{\rm cycle}$; and the total proton confinement time $\tau_{\rm conf}$: (i) $\tau_{\rm int} \gg \tau_{\rm cycle}$; (ii) $\tau_n > \tau_{\rm cycle}$; (iii) $\tau_{\rm int} \ll \tau_{\rm conf}$. The first condition ensures that the protons attain sufficient energy. Conditions (i) and (ii) allow the neutrons to escape the source before decaying. Condition (iii) permits sufficient interaction to produce neutrons and neutrinos. These three conditions together define an optically thin source [18].

If CR sources in the Galaxy are optically thin, then one would expect a flux of neutrinos with a spectrum $\propto E^{-2.34}$. If the maximum attainable energy of cosmic rays in the Galaxy is $E_{\rm max} \sim 1 - 3$ EeV, by Hillas criterion [19] we would expect protons to be accelerated to $E^p_{\rm max} = E_{\rm max}/26 \sim 40$ PeV and therefore secondary neutrinos to be be produced with energies of up to 2 to 3 PeV.

Quite recently the IceCube Collaboration has reported a preliminary observation of 26 neutrino candidates [20], in addition to but at lower energies than the two ~ 1 PeV neutrinos reported earlier in 2013 [21]. Taken together, these 28 events constitute a 4.1σ excess compared to expectations from background. Interpretation of these results, however, does not appear to be entirely straightforward. For instance, if one makes the common assumption of an unbroken E_ν^{-2} neutrino energy spectrum, then one expects to observe about 8-9 events with higher energies than the two highest energy events observed thus far. Recently, we have investigated the compatibility between the IceCube observations and the hypothesis of an unbroken power-law spectrum arising from optically thin Galactic neutrino sources, finding that a cosmic neutrino flux per flavor, averaged over all three flavors,

$$\frac{dF_\nu}{d\Omega dA dt dE_\nu} = 6.62 \times 10^{-7} \left(\frac{E_\nu}{1 \text{ GeV}}\right)^{-2.3} (\text{GeV} \cdot \text{cm}^2 \cdot \text{s} \cdot \text{sr})^{-1}, \quad (3)$$

is consistent with the data reported thus far [22].

The assumption underlying the leaky box model is that the energy density in CRs observed locally is typical of other regions of the Galactic disk. If so, the total power required to maintain the cosmic radiation in equilibrium can be obtained by integrating the generation rate of primary

Journal of Physics: Conference Series **531** (2014) 012009 doi:10.1088/1742-6596/531/1/012009

CRs over energy and space. Using (2), we obtain

$$\frac{d\epsilon_{\rm CR}}{dt} = \int d^3x \int Q(E)\, dE = V_G \frac{4\pi}{c} \int \frac{J(E)}{\tau(E/Z)} dE\,, \tag{4}$$

where $V_G \sim 10^{67}$ cm^3 is the Galactic disk volume [24]. For $E_{\rm knee} < E < E_{\rm ankle}$, we conservatively assume that the trapping time in the Galaxy scales with energy as $\tau = 2 \times 10^7 (E_{\rm GeV}/Z)^{-1/3}$ yr [25]. In this case the power budget required to fill in the spectrum from the knee to the ankle is found to be $d\epsilon_{\rm CR}/dt \simeq 2 \times 10^{39}$ erg/s [25]. From this, we deduced elsewhere [22] the energy transfer fraction from the parent protons to the pions which ultimately produce the observed neutrinos, demonstrating that pp collisions are more likely to produce the neutrino flux than are $p\gamma$ collisions, and indeed that the power budget is enough to reproduce the neutrino flux in (3).

It is interesting to employ existing limits on high energy photons to check the plausibility of our hypothesis that the IceCube excess is of Galactic origin. γ rays are produced by π^0 decays at the same optically thin sources where neutrinos are produced by $\pi^\pm$ decay. As described in [13], one can predict a differential γ ray flux based the best-fit single power law ν flux discussed in this paper, and compare to measurements. The CASA-MIA 90% C.L. upper limits on the integral diffuse γ ray flux, I_γ for energy bins,

$$\frac{E_\gamma^{\rm min}}{\rm GeV} = 3.30 \times 10^5,\ 7.75 \times 10^5,\ 2.450 \times 10^6\,, \tag{5}$$

are

$$\frac{I_\gamma}{\rm cm^{-2}\ s^{-1}\ sr^{-1}} < 1.0 \times 10^{-13},\ 2.6 \times 10^{-14}, 2.1 \times 10^{-15}\,, \tag{6}$$

respectively [26]. A more rigorous comparison would involve measurements on the diffuse γ fay flux within about 15° of the Galactic plane. The CASA-MIA Collaboration has in fact studied γ ray emission from the direction of the Galactic plane, reporting the flux limits as a fraction of the CR flux [27] rather than an integral bound. Comparing the relative fractions from the all-sky analysis to the Galactic plane analysis indicates that constraining the observation to the Galactic plane region does indeed lead to tighter constraints:

$$\frac{I_\gamma}{\rm cm^{-2}\ s^{-1}\ sr^{-1}} \lesssim 5.0 \times 10^{-14},\ 1.3 \times 10^{-14},\ 2.1 \times 10^{-15}\,, \tag{7}$$

at the 90% C.L. Under the simplifying assumption that there is no photon absorption, the integral photon fluxes we predict based on our single power law hypothesis (in units of photons cm^{-2} s^{-1} sr^{-1}), above the energies specified in (5), are

$$\int_{E_\gamma^{\rm min}} \frac{dF_\gamma}{d\Omega dA dt dE_\gamma} dE_\gamma = 4.2 \times 10^{-14},\ 1.4 \times 10^{-14},\ 3.1 \times 10^{-15}\,. \tag{8}$$

For the first two energy bins, the predicted fluxes saturate the 90% C.L. measurements of CASA-MIA, while the last bin slightly exceeds the 90% C.L. bound. This does not, however, imply that the Galactic origin hypothesis is ruled out at 90% C.L. First of all, one must keep in mind that sources which are optically thin up to $E_\gamma \sim 100$ TeV may not be optically thin at higher energies, suggesting that the importance of photon bounds in establishing the origin of the IceCube excess should be considered with some caution. Even if we ignore this caveat, we still do not know the maximum neutrino energy reached at acceleration sites, so the maximum photon energy is likewise unknown. In addition, absorption becomes important in the energy regime covered by the last bin, as mean free path of PeV photons in the CMB is about 10 kpc.

Note that $R_G \sim 10$ kpc, leading to an interesting signature: Photons coming from "our half" of the Galaxy will be largely unattenuated, while those from the farther half will be significantly attenuated. Since both photons and neutrinos point back to the sources, coordinated comparisons of neutrino and photon data will facilitate a completely new exploration of the highest-energy Galactic sources. As described in [13], taking into account absorption of the photon flux for $E_\gamma^{\rm min} > 1$ PeV leads to about a 12% reduction in the predicted photon flux. Furthermore, varying the photon maximum energy cutoff of Eq. (8) to,

$$\frac{E_\gamma^{\rm max}}{\rm PeV} = 6,\ 7,\ 8\,, \tag{9}$$

we obtain

$$\int_{E_\gamma^{\rm min}}^{E_\gamma^{\rm max}} \frac{dF_\gamma}{d\Omega dA dt dE_\gamma} dE_\gamma = 2.1 \times 10^{-15},\ 2.3 \times 10^{-15},\ 2.4 \times 10^{-15}\,. \tag{10}$$

From the discussion above, we can see there are several ways to comply with the CASA-MIA bound. For instance, $E_\gamma^{\rm max} = 6$ PeV is already consistent with the measured bound, even without absorption. For higher energies, absorption provides enough reduction of the photon flux to retain consistency with measurements. It is also worth noting that the comparison discussed here is based on experimental bounds on the all-sky γ ray flux.

Only the IceCube Collaboration has thus far reported constraints on γ ray emission between 1 and 10 PeV from the direction of the Galactic plane. Bounds from the IceCube 40 string configuration [30], are not restrictive enough to challenge the Galactic origin hypothesis. However, within 5 years of data taking with the complete IceCube configuration of 86 strings, enough statistics will be gathered to elucidate the $\nu - \gamma$ ray connections.

Finally we comment on the consistency between the arrival direction distribution of the IceCube excess and the hypothesis that the sources are nearby. Fourteen of the 26 reported neutrino events arrive from within about 15° of the Galactic plane, including one of the two highest energy events, which coincides with the Galactic center (within errors) [28] . The highest energy event is outside of this angular window, but (as noted in [29]) does correspond with a possible hotspot in the IceCube photon search [30]. This could reflect emission of neutrinos and γ rays from a common, nearby source, as γ rays do not survive propagation further than ~ 10 kpc. The recently discovered large reservoir of ionized gas extending over a large region around the Milky Way [31] could provide the target material required for neutrino production outside the Galactic disk in models in which proton diffusion extends to the Galactic halo [32, 33]. However, given the current statistics and the insufficient understanding of the atmospheric (in particular the prompt neutrino [34]) background, the arrival direction distribution neither favors nor disfavors a Galactic origin [29, 35]. More data are required to settle the issue.

Acknowledgments
This work was supported by the US NSF grant numbers: CAREER PHY1053663 (LAA); PHY-0757959 (HG); PHY-1205854 (TCP); PHY-1068696, PHY- 1125897 (AVO); US DoE grant DE-FG05-85ER40226 (TJW); NASA grant numbers: NNX13AH52G (LAA, TCP), 11-APRA- 0066 (AVO); the Simons Foundation (grant #306329) (TJW); and the UWM RGI (BJV).

References

[1] R. Abbasi *et al.* [HiRes Collaboration], Phys. Rev. Lett. **100**, 101101 (2008) [arXiv:astro-ph/0703099]; J. Abraham *et al.* [Pierre Auger Collaboration], Phys. Rev. Lett. **101**, 061101 (2008) [arXiv:0806.4302].
[2] J. R. Hörandel, Astropart. Phys. **19**, 193 (2003) [arXiv:astro-ph/0210453]; P. Blasi and E. Amato, JCAP **1201**, 010 (2012) [arXiv:1105.4521].
[3] T. Abu-Zayyad *et al.*, Astrophys. J. **557**, 686 (2001) [arXiv:astro-ph/0010652].
[4] D. J. Bird *et al.* [HIRES Collaboration], Phys. Rev. Lett. **71**, 3401 (1993).

[5] R. Abbasi *et al.* [IceCube Collaboration], Astrophys. J. **746**, 33 (2012) [arXiv:1109.1017 [hep-ex]].
[6] J. Linsley, Phys. Rev. Lett. **34**, 1530 (1975).
[7] J. Aublin and E. Parizot, Astron. Astrophys. **441**, 407 (2005) [astro-ph/0504575].
[8] M. Aglietta *et al.* [EAS-TOP Collaboration], Astrophys. J. **692**, L130 (2009) [arXiv:0901.2740 [astro-ph.HE]].
[9] A. Aab *et al.* [Pierre Auger Collaboration], arXiv:1307.5059 [astro-ph.HE]. For previous analyses, see also, P. Abreu *et al.* [Pierre Auger Collaboration], Astrophys. J. Suppl. **203**, 34 (2012) [arXiv:1210.3736 [astro-ph.HE]]; P. Abreu *et al.* [Pierre Auger Collaboration], Astropart. Phys. **34**, 627 (2011) [arXiv:1103.2721 [astro-ph.HE]]. A summary of the Auger results has been preseneted in these proceedings, O. Deligny [for the Pierre Auger Collaboration], arXiv:1403.5569 [astro-ph.HE].
[10] N. Hayashida *et al.* [AGASA Collaboration], Astropart. Phys. **10**, 303 (1999) [astro-ph/9807045].
[11] T. Antoni *et al.* [KASCADE Collaboration], Astrophys. J. **604**, 687 (2004) [astro-ph/0312375].
[12] M. Stumpert, PhD Thesis, FZKA-7380.
[13] L. A. Anchordoqui, V. Barger, I. Cholis, H. Goldberg, D. Hooper, A. Kusenko, J. G. Learned, D. Marfatia, S. Pakvasa, Thomas C. Paul, and Thomas J. Weiler, JHEAp **1**, 1 (2014) [arXiv:1312.6587 [astro-ph.HE]].
[14] M. Gupta and W. R. Webber, Astrophys. J. **340**, 1124 (1989); S. P. Swordy *et al.*, Astrophys. J. **403**, 658 (1993).
[15] P. Blasi and E. Amato, JCAP **1201**, 011 (2012).
[16] P. L. Biermann Astron. Astrophys. **271**, 649 (1993); P. L. Biermann, T. K. Gaisser and T. Stanev, Phys. Rev. D **51**, 3450 (1995); A. M. Hillas, J. Phys. G **31**, R95 (2005).
[17] S. I. Syrovatskii, Comment. Astrophys. Space Phys. **3**, 155 (1971); V. S. Ptuskin *et al.*, Astron. Astrophys. **268**, 726 (1993); J. Candia, S. Mollerach and E. Roulet, JCAP **0305**, 003 (2003).
[18] M. Ahlers, L. A. Anchordoqui, H. Goldberg, F. Halzen, A. Ringwald and T. J. Weiler, Phys. Rev. D **72**, 023001 (2005) [astro-ph/0503229].
[19] The Hillas criterion states that the maximum attainable energy is limited by the fact that particles can only be accelerated in a region until their Larmor radius exceeds its size. A. M. Hillas, Ann. Rev. Astron. Astrophys. **22**, 425 (1984).
[20] M. G. Aartsen *et al.* [IceCube Collaboration], Science **342**, no. 6161, 1242856 (2013) [arXiv:1311.5238 [astro-ph.HE]]; F. Halzen, arXiv:1308.3171.
[21] M. G. Aartsen *et al.* [IceCube Collaboration], Phys. Rev. Lett. **111**, 021103 (2013) [arXiv:1304.5356].
[22] L. A. Anchordoqui, H. Goldberg, M. H. Lynch, A. V. Olinto, T. C. Paul and T. J. Weiler, Phys. Rev. D, in press [arXiv:1306.5021 [astro-ph.HE]].
[23] T. K. Gaisser, F. Halzen and T. Stanev, Phys. Rept. **258**, 173 (1995) [Erratum-ibid. **271**, 355 (1996)] [arXiv:hep-ph/9410384].
[24] T. K. Gaisser, arXiv:astro-ph/0501195.
[25] T. K. Gaisser, J. Phys. Conf. Ser. **47**, 15 (2006).
[26] M. C. Chantell *et al.* [CASA-MIA Collaboration], Phys. Rev. Lett. **79**, 1805 (1997) [astro-ph/9705246].
[27] A. Borione *et al.*, Astrophys. J. **493**, 175 (1998) [astro-ph/9703063].
[28] S. Razzaque, Phys. Rev. D **88**, 081302 (2013) [arXiv:1309.2756 [astro-ph.HE]].
[29] M. Ahlers and K. Murase, arXiv:1309.4077 [astro-ph.HE].
[30] M. G. Aartsen *et al.* [IceCube Collaboration], Phys. Rev. D **87**, no. 6, 062002 (2013) [arXiv:1210.7992 [astro-ph.HE]].
[31] A. Gupta, S. Mathur, Y. Krongold, F. Nicastro and M. Galeazzi, Astrophys. J. **756**, L8 (2012) [arXiv:1205.5037 [astro-ph.HE]].
[32] F. C. Jones, A. Lukasiak, V. Ptuskin and W. Webber, astro-ph/0007293. V. S. Ptuskin, O. N. Strelnikova and L. G. Sveshnikova, Astropart. Phys. **31**, 284 (2009). V. Ptuskin, Astropart. Phys. **39-40**, 44 (2012).
[33] Precise calculations for neutrino production in the Galactic halo have been recently carried out, A. M. Taylor, S. Gabici and F. Aharonian, arXiv:1403.3206 [astro-ph.HE].
[34] P. Lipari, arXiv:1308.2086 [astro-ph.HE].
[35] A. Neronov, D. V. Semikoz and C. Tchernin, arXiv:1307.2158 [astro-ph.HE].

Journal of Physics: Conference Series **531** (2014) 012010 doi:10.1088/1742-6596/531/1/012010

Anisotropies in TeV Cosmic Rays Related to the IBEX Ribbon

N. A. Schwadron[1,2], F. C. Adams[3], E. Christian[4], P. Desiati[5], P. Frisch[6], H. O. Funsten[7], J. R. Jokipii[8], D. J. McComas[2,9], E. Moebius[1], G. Zank[10]

[1] University of New Hampshire, Durham, NH, 03824
[2] Southwest Research Institute, San Antonio, TX 78228, USA
[3] University of Michigan, Ann Arbor, MI, 48109
[4] Goddard Spaceflight Institute
[5] IceCube Research Center and Astronomy Department, University of Wisconsin, Madison, WI 53706
[6] University of Chicago, Department of Astronomy and Astrophysics, Chicago, IL 60637
[7] Los Alamos National Laboratory, Los Alamos, NM, 87545
[8] University of Arizona
[9] University of Texas, San Antonio, TX 78228, USA
[10] University of Alabama Huntsville

E-mail: n.schwadron@unh.edu

Abstract. The Interstellar Boundary Explorer (IBEX) observes enhanced Energetic Neutral Atoms (ENAs) emission in the keV energy range from a narrow ($\sim$20° wide) "ribbon" in the sky that appears to be centered on the direction of the local interstellar (LIS) magnetic field. The Milagro collaboration, the Asγ collaboration and the IceCube observatory have recently made global maps of cosmic ray fluxes in the TeV energy range, revealing anisotropic structures ordered in part by the local interstellar magnetic field and the interstellar flow. This paper following from a recent publication in *Science* makes the link between these disparate observations by developing a simple model of the magnetic structure surrounding the heliosphere in the Local Interstellar Medium (LISM) that is consistent with both IBEX ENA fluxes and TeV cosmic ray anisotropies. The model also employs the revised velocity direction of the LIC derived from neutral He observations by IBEX. By modeling the propagation of cosmic rays through this magnetic field structure, we specifically show that (1) the large-scale TeV anisotropy provides a roughly consistent orientation for the local interstellar magnetic field at the center of the IBEX Ribbon and corroborates the $\sim 3\ \mu$G magnitude of the local interstellar magnetic field derived from IBEX observations of the global heliosphere; (2) and small-scale structures in cosmic rays (over $< 30°$ angular scales) are influenced by the interstellar field interaction with the heliosphere at energies < 10 TeV. Thus, we provide a link between IBEX ENA observations, IBEX neutral observations of interstellar He, and TeV cosmic ray anisotropies, which are strongly influenced by the interactions between the local interstellar magnetic field, the flow of the local interstellar plasma, and the global heliosphere.

1. Introduction

The Interstellar Boundary Explorer Mission (IBEX) was launched October 19, 2008, with the objective to discover the global interaction between the solar wind and the local interstellar medium [1]. IBEX measures Energetic Neutral Atoms (ENAs) created predominantly from charge-exchange between neutral interstellar hydrogen and plasma protons in the solar wind and interstellar medium [1; 2]. First results from IBEX include global maps of the heliosphere in ENAs [3–6] and the first measurements of low energy interstellar oxygen, helium and hydrogen atoms [7]. Higher energy maps from INCA were also revealed [8]. The IBEX and INCA sky maps showed striking differences between observations and predictions from models [6].

The IBEX maps revealed the existence of a narrow ribbon ($\sim 20°$ wide) of elevated ENA emissions [3–5] that forms a circular structure centered on (RA, DEC) $\sim (48.5°\pm1.5, -21.2°\pm1.6)$, which is likely near the local interstellar (LISM) magnetic field direction [6].

This paper discusses another observational signature of this LISM magnetic field in TeV cosmic ray anistropies [9]. Figure 1 shows the cosmic ray gyroradius for a typical field strength in the heliosheath (1 μG, blue) and LISM (3 μG, red). In the energy range from 0.5 to 10 TeV, cosmic rays with gyroradii of 100's of AU probe the scales in the LISM that are larger than the heliosphere. Recall that the deformation of the LIsMF also extends to 100's of AU. Therefore, it is precisely in the $\sim$ 0.5-10 TeV energy range where cosmic rays may probe the same large scale LIsMF structures apparently manifested in the IBEX Ribbon.

2. Cosmic Ray Anisotropy Observations

The arrival direction distribution of cosmic rays from tens of GeV to tens of TeV is found to have an asymmetry which is not compatible with the convection anisotropy expected by the motion of the solar system through the galactic medium [10; 11]. A broad relative excess is observed between equatorial right ascension $\alpha \approx 18$ hr and $\alpha \approx 8$ hr, with an amplitude of $10^{-4} - 10^{-3}$. The global structure of the anisotropy depends on the cosmic ray energy, but is found to change weakly in the energy range from about 1 TeV to a few tens of TeV. Fig. 2 (panel A) shows the map in equatorial coordinates of relative intensity in the cosmic ray arrival direction, as observed by Tibet ASγ in the northern equatorial hemisphere at a median energy of about 5 TeV [12], and by the IceCube Observatory in the southern equatorial hemisphere at a median energy of about 20 TeV [13]. In each declination band of about 5° the relative intensity represents the modulation of the counting rate with respect to the mean rate in that declination. The two observations reported in the figure correspond to different median energies, although the observations span a wide energy range to collect the large event sample needed for the determination of the small anisotropy amplitude. In particular IceCube reported that 68% of the data sample with median energy of about 20 TeV is between 4 TeV and 63 TeV [14]. Assuming an equivalent resolution for the Tibet ASγ, it turns out that the two data samples are significantly overlapped in cosmic ray energy, and therefore they can be directly compared. In Fig. 2 (panel A) it appears that the relative excess in arrival direction is oriented toward the region of the sky that comprises the downstream direction of the interstellar wind and the downfield direction for the LIsMF.

In order to accentuate the small scale features that appear on the global anisotropy sky map, the Milagro Collaboration subtracted a 30°-averaged sky map from that of the global anisotropy [see 15, for details]. The residual map highlights all the fine structures with characteristic angular scales that are smaller than 30°. A similar analysis was performed by the IceCube Observatory, where the averaging procedure was performed within 60°, meaning that only features smaller than such an angular scale are highlighted. Fig. 2 (panel B) shows the corresponding map in equatorial coordinates of the statistical significance with respect to the averaged maps and with a 10° smoothing. The northern hemisphere shows the result from Milagro at a mean energy of $\sim$1 TeV and the southern hemisphere shows the IceCube observations at a median energy of $\sim$ 20 TeV. Once again, although the characteristic energies of the two observations are different, the energy ranges of the observations significantly overlap. Some interesting visual correlation between the two halves of the sky arise.

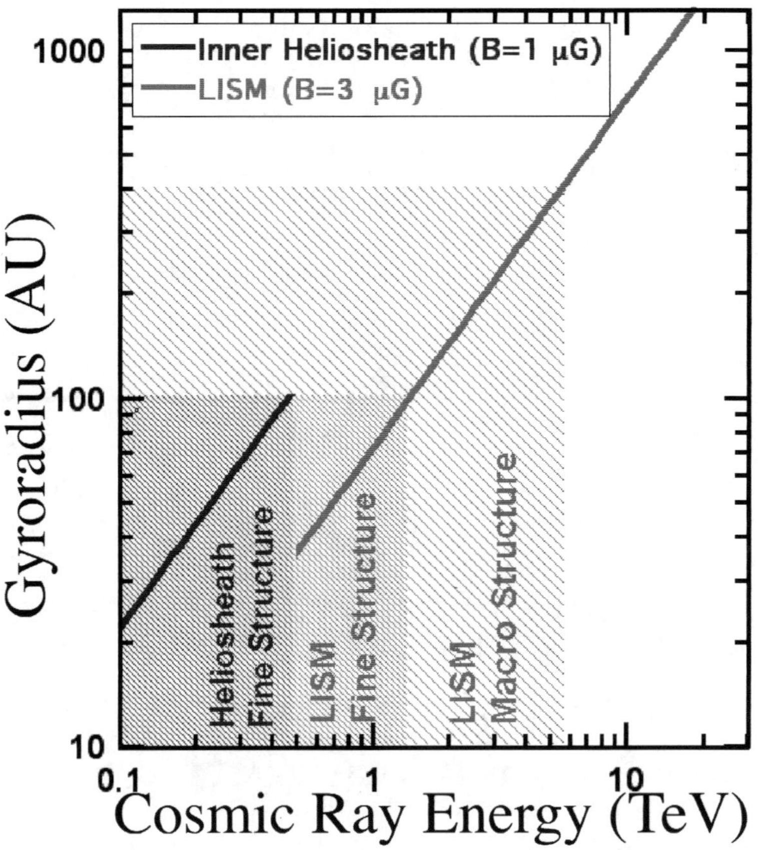

Figure 1. The gyroradius of cosmic rays in the heliosheath (1 μG, blue) and LISM (3 μG, red) probe increasingly large structures with increasing energy. Cosmic rays probe the heliosheath structure in the $\sim$ 0.1-0.5 TeV energy range, LISM fine structure < 100 AU in the $0.5 - 1.5$ TeV energy and LISM macrostructure including the distortion of the LISM field around the heliosphere of 100's of AU for energies between $\sim 1.5 - 10$ TeV. It is over this energy TeV energy range where cosmic rays probe the same magnetic structures from the LISM that likely produce the IBEX Ribbon. [From 9].

In both sky maps, the localized excess of cosmic ray flux observed at equatorial coordinates $(\alpha, \delta) \approx$ (21 hr, +32°) is due to gamma rays from the Cygnus region to which both Tibet ASγ and Milagro are sensitive. In the southern hemisphere, the most significant localized excess and deficit regions at $(\alpha, \delta) \approx$ (7 hr, -45°) and $(\alpha, \delta) \approx$ (15 hr, -45°), respectively, correspond to the center of the global anisotropy regions, meaning that some of the large scale structure influences features in the residual sky map. The Milagro observation of a localized fractional excess in the direction of the heliotail does not appear to have a correspondance in the southern sky. On the other hand, the filamentary feature at right ascension $\alpha \approx$ 8-10 hr, appears to continue throughout the South in a quasi-circular pattern, even though here its statistical significance is not as high. Given the width of the energy response functions, it is possible that cosmic rays contributing to different localized regions have different characteristic energy. More observations are needed to refine these maps.

Fig. 2 (panel C) shows the IBEX Ribbon observed at 1.1 keV [3–6] in J2000 coordinates. We include labels for the upwind and downwind directions [16], the Voyager 1 (V1) and Voyager 2

(V2) directions and the upfield and downfield directions associated with the unperturbed LISM magnetic field. The black line shows the magnetic equator of the external LIsMF in the direction of the IBEX Ribbon center. The shift of the Ribbon to the inside of the LIsMF magnetic equator is likely due to flaring of the field on the noseward side of the heliosphere [6]. The global structure in all three panels is ordered about the LIsMF magnetic equator, which reinforces the concept that the LIsMF influences both TeV cosmic rays and the ENA fluxes observed by IBEX.

3. Local Interstellar Structure

The heliosphere interacts with the local interstellar flow that originated from the center of Loop I. Typical temperatures are $\sim 3000 - 13000$ K and proton densities $\sim 0.05 - 0.1$ cm^{-3} [18]. Deviations of the velocities of individual clouds from a rigid-body flow indicate that the flow is decelerating [19]. A decelerating flow sets the conditions for collisions between the local clouds. The Sun is inside of the LIC and no more than 0.1 pc away from the LIC edge in the direction of the galactic center hemisphere [20; 21]. In the direction of the anticenter hemisphere, the closest cloud to the LIC is the 'Blue Cloud' [BC, 22; 23]. Linsky *et al.* [24] show that the conditions for colliding clouds exist within $\sim$ 10 pc, with relative velocities between overlapping clouds of up to 50 km s^{-1} (Figure 3). Among the closest of possible collisions would be a collision between the LIC and BC.

Cloud collisions in this region are supported by observations of short-timescale scintillation towards two compact radio sources, PKS 0405-385 and PSR J0437-47, which are located behind the BC [25]. Models of the timescales and frequency dependence of the scintillation show that the scattering screens, consisting of partially ionized gas such as the LIC, are within $\sim$ 10 pc. Towards PSR J0437-47, the local turbulence occurs over spatial scales of $\sim$ 0.02 AU [26]. PKS 0405-385 and PSR J0437-47 are located within 29° of the downfield direction.

The decelerating flow of local interstellar clouds past the Sun appears to be ordered by an evolved superbubble associated with Loop I [18]. An interstellar magnetic field interacting with the compressed, perhaps inhomogeneous, shell provides a natural explanation for inhomogeneities in the LIsMF. Figure 4 shows the flow of local clouds through the LSR for a meridianal cut through the galactic plane where the Z-direction is directed toward galactic north and the X-direction is toward galactic center. Measurements of interstellar He I inside of the heliosphere provide the LIC velocity [27; 28], while velocities for 15 local clouds have been determined with the triangulation of the radial velocity component towards different stars behind the clouds [25]. The Loop I superbubble shell is represented by the thick arc, which is based on the S1 shell of Wolleben [29] that is centered $\approx$ 78 pc away at galactic coordinates of L,B$\approx$346°,3° [29].

Measurements of He I inside of the heliosphere provide the LIC velocity [27; 28]. Pressure equilibrium between the gas and magnetic field in the LIC give a LIC magnetic field strength of $\sim 2.7\ \mu$G, which is comparable to the $\sim 3\ \mu$G field strength obtained from the globally distributed ENA fluxes [30].

4. Imprint of the heliosphere on the global anisotropy in TeV cosmic rays

Schwadron *et al.* [9] utilized the observations of IBEX to estimate the direction of the magnetic field in the LISM. Knowing in addition the flow velocity of the LISM based on neutral He measurements, we developed model estimates for the global anisotropy of TeV cosmic rays in the LISM associated with modulation by the local interstellar flow. The flow configuration of the LISM has a flow velocity almost perpendicular to the local interstellar magnetic field and consistent with the outward expansion of plasma from the Loop I S1 superbubble (Fig. 5). The model also includes the propagation effects associated with deformation of the local interstellar magnetic field about the heliosphere. The resulting global anisotropy map in TeV cosmic rays is shown in Fig. 6. The full calculation is detailed by Schwadron et al. [9]. The main free parameter

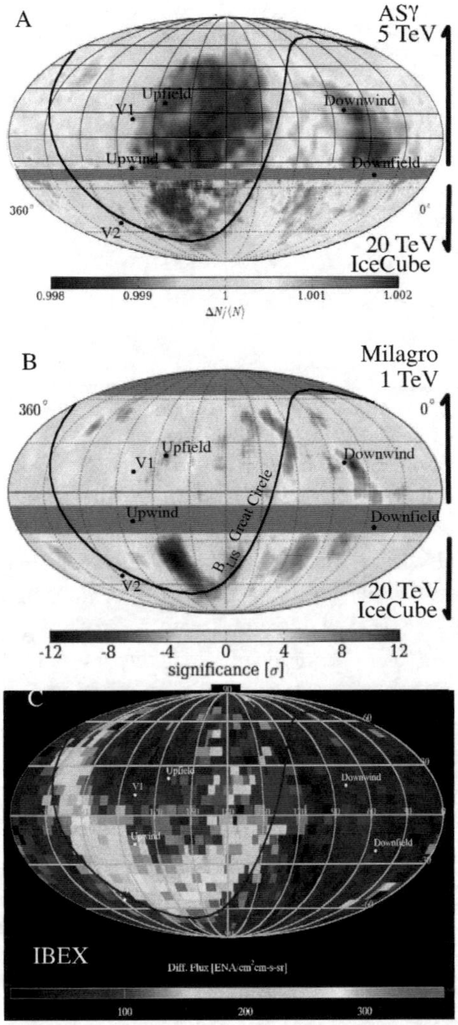

Figure 2. *Panel A*: relative intensity observed by Tibet Asγ collaboration at a median energy of ~ 5 TeV [17] in the north and by IceCube in the south at a median energy of ~ 20 TeV [13]. Both portions of the maps are smoothed with 3°-5°. *Panel B*: significance map that smoothes (over 30°-60°) and filters out the large scale anisotropy to reveal only the small scale anisotropy. The northern hemisphere is from the Milagro project, centered on ~ 1 TeV median energy [15]. The southern hemisphere is the map from the IceCube collaboration at $\sim$20 TeV median energy [13]. *Panel C*: The IBEX Ribbon observed at ~ 1.1 keV [3–6] in J2000 coordinates and the TeV Milagro/Ice Cube data. We include labels for the upwind and downwind directions [16], the Voyager 1 (V1) and Voyager 2 (V2) directions and the upfield and downfield directions associated with the unperturbed LIsMF. The TeV cosmic ray data appears to be ordered by the LISM magnetic equator. These plots are in equatorial coordinates with $\alpha = 0$ hr at the right, and increasing towards the left.

in these estimates of the global TeV anisotropy is the level of perpedindicular diffusion for which we use typical parameters of the ISM that $\kappa_\perp/\kappa_\parallel \sim 0.003$ [32].

Supernova remants (SNR) also contribute to the anisotropy, at least in localized regions. While the direction of LISM magnetic field found by IBEX must play a significant role in

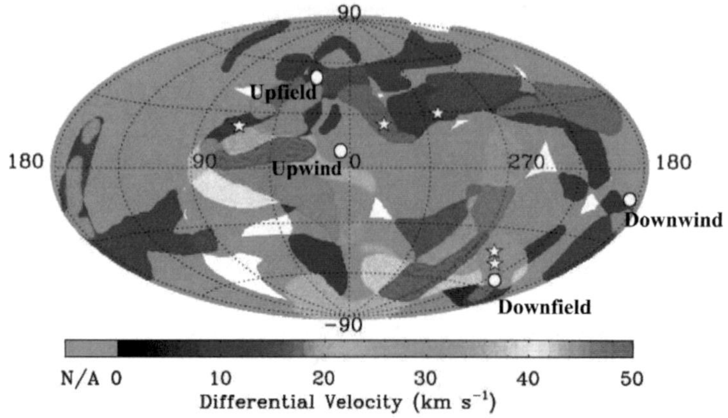

Figure 3. Zones of LISM cloud interaction with superposed upwind, downwind, upfield and downfield directions. Colored regions show the directions in which multiple LISM clouds are detected along the line of sight. The white regions indicate where no LISM cloud is detected, and gray regions where only a single cloud is detected. The color coding signifies the maximum magnitude of the differential velocity between clouds along the same line of sight. Five scintillation sight lines are shown here by star symbols. This plot is in galactic coordinates, and centered on the galactic center with longitude increasing towards the left. [Based on 24].

ordering the cosmic ray anisotropy map, the sign of the dipolar anisotropy component could either align with or oppose the component that arises from local interstellar modulation. Further, the magnitude of the dipolar component of the cosmic ray anisotropy is poorly constrained by observations.

5. Conclusions

The comparison shown in Figure 5 confirms the direction of the LISM magnetic field from the center of the IBEX ribbon. The result is somewhat surprising since the LISM magnetic field from the IBEX ribbon is determined in a very local interstellar environment, within 100's of AU from the heliosphere. Turbulence in the LISM might be expected to destroy signatures of the direction inferred from IBEX over 1-10 pc correlation scales. Such effects would obscure any relationship between global TeV cosmic ray anistropies and the IBEX ribbon. The fact that a relationship is observed suggests that the LISM magnetic field remains relatively coherent over scales of at least 1-10 pc. This in turn suggests that our heliosphere is currently moving through a magnetically ordered portion of the LISM. This reinforces the concept that the LISM magnetic field has sufficient pressure to play a significant role in LISM dynamics.

Thus, we have shown that TeV cosmic rays provide independent confirmation of the direction of the LISM magnetic field inferred from the center of the IBEX ribbon. The correlation is remarkable given that these independent observations are taken over vastly different energy ranges. IBEX measures energetic neutral atoms at thousands of eV and TeV anisotropy maps are formed from particles at higher energies by ten orders of magnitude. Our increasing understanding of the local interstellar environment is particularly important considering that the Voyager 1 spacecraft is now entering interstellar space.

5.1. Acknowledgments

We are deeply indebted to all of the outstanding people who have made the IBEX mission possible and have contributed to the Asγ, IceCube and Milagro projects. This work was carried

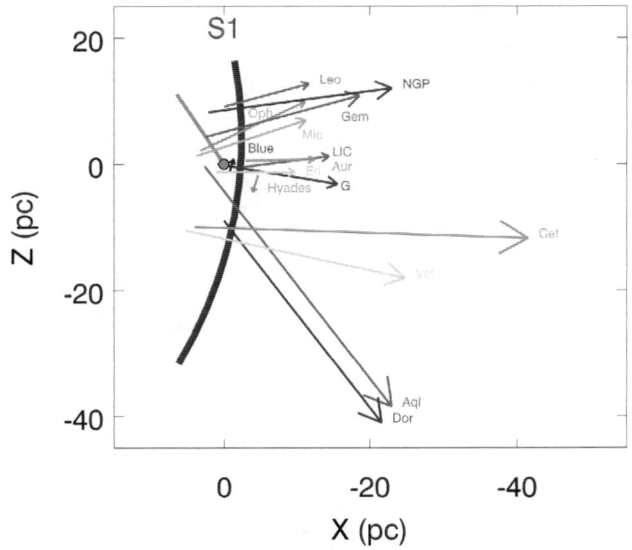

Figure 4. The arrows show the velocities of local interstellar clouds through the LSR in the X-Z meridian plane, where the X-axis is directed towards the galactic center and the Z-axis is directed towards the galactic north pole [based on heliocentric velocities in 25]. The LSR velocities are calculated using the solar apex motion [from 31]. The arrow origin is located at the cloud-center and the distance of the nearest star (that is not the Sun) in the cloud. Arrows are labeled with cloud names. The thick gray arc shows the location of the Loop I S1 superbubble, based on a thin shell with radius of 78 pc that is centered at galactic coordinates of L,B=346°,3° [29]. The purple line gives the direction of the LIsMF [4] and the orange dot shows the solar location.

out as a part of the IBEX project, with support from NASAs Explorer Program.

References

[1] McComas D J, Allegrini F, Bochsler P, Bzowski M, Collier M, Fahr H, Fichtner H, Frisch P, Funsten H O, Fuselier S A, Gloeckler G, Gruntman M, Izmodenov V, Knappenberger P, Lee M, Livi S, Mitchell D, Möbius E, Moore T, Pope S, Reisenfeld D, Roelof E, Scherrer J, Schwadron N, Tyler R, Wieser M, Witte M, Wurz P and Zank G 2009 *Space Sci. Rev.* **146** 11–33

[2] McComas D J, Funsten H O, Fuselier S A, Lewis W S, Möbius E and Schwadron N A 2011 *Geophys. Res. Lett.* **38** L18101

[3] McComas D J, Allegrini F, Bochsler P, Bzowski M, Christian E R, Crew G B, DeMajistre R, Fahr H, Fichtner H, Frisch P C, Funsten H O, Fuselier S A, Gloeckler G, Gruntman M, Heerikhuisen J, Izmodenov V, Janzen P, Knappenberger P, Krimigis S, Kucharek H, Lee M, Livadiotis G, Livi S, MacDowall R J, Mitchell D, Möbius E, Moore T, Pogorelov N V, Reisenfeld D, Roelof E, Saul L, Schwadron N A, Valek P W, Vanderspek R, Wurz P and Zank G P 2009 *Science* **326** 959–

[4] Funsten H O, Allegrini F, Crew G B, DeMajistre R, Frisch P C, Fuselier S A, Gruntman M, Janzen P, McComas D J, Möbius E, Randol B, Reisenfeld D B, Roelof E C and Schwadron N A 2009 *Science* **326** 964–

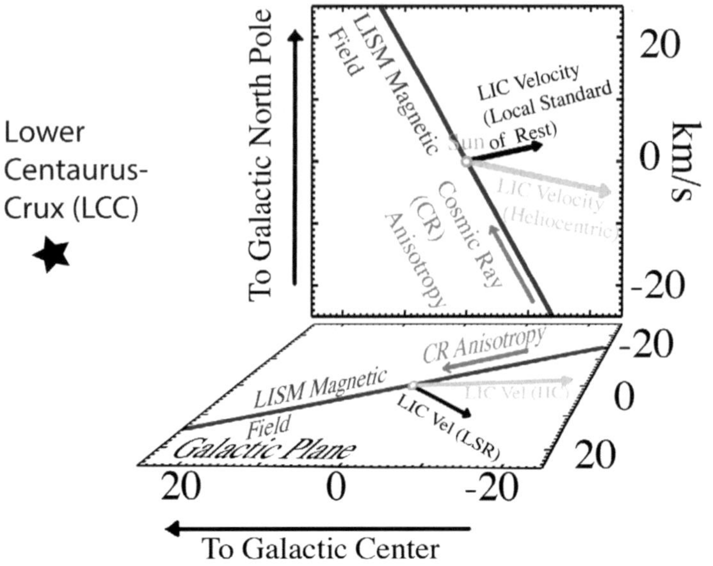

Figure 5. Directions and magnitudes of the LIC velocities inferred from IBEX in the heliocentric (HC; green vector) rest frame and the local standard of rest (LSR; black vector) in the galaxy. These vectors are shown in the galactic plane (lower plane) and the plane containing the galactic poles (upper plane) together with the magnetic field of the local interstellar medium (blue lines). The LIC velocity and magnetic field are a part of the Loop I superbubble roughly centered on the Lower Centaurus Crux (LCC). The cosmic ray anisotropy (grey vector) indicates average streaming of cosmic rays.

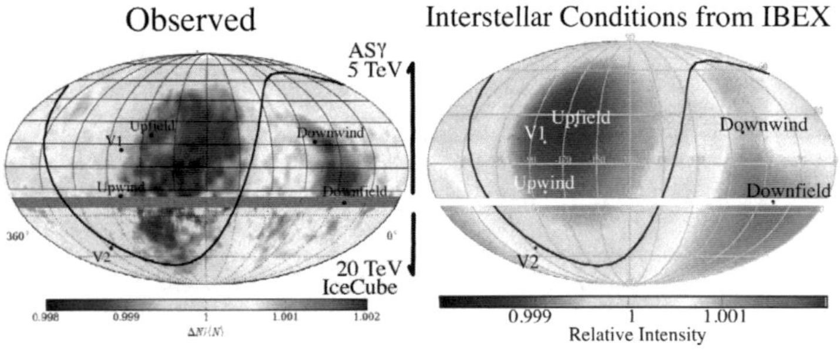

Figure 6. Comparison between observed (left) and modeled (right) cosmic ray intensities across the sky (J2000 coordinates). Black curves show the magnetic equator given a magnetic field direction derived from the center of the IBEX ribbon. (Left) The regions below 25° S latitude show the anisotropy map from IceCube with a median energy of 20 TeV and above 20° S latitude is the anisotropy map from AS-γ with 5 TeV median energy. (Right) The modeled map at 20 TeV is shown below 25° S latitude and at 5 TeV above 20° S latitude. [From 9].

[5] Fuselier S A, Allegrini F, Funsten H O, Ghielmetti A G, Heirtzler D, Kucharek H, Lennartsson O W, McComas D J, Möbius E, Moore T E, Petrinec S M, Saul L A, Scheer J A, Schwadron N and Wurz P 2009 *Science* **326** 962

[6] Schwadron N A, Bzowski M, Crew G B, Gruntman M, Fahr H, Fichtner H, Frisch P C, Funsten H O, Fuselier S, Heerikhuisen J, Izmodenov V, Kucharek H, Lee M, Livadiotis G, McComas D J, Moebius E, Moore T, Mukherjee J, Pogorelov N V, Prested C, Reisenfeld D, Roelof E and Zank G P 2009 *Science* **326** 966–

[7] Möbius E, Bochsler P, Bzowski M, Crew G B, Funsten H O, Fuselier S A, Ghielmetti A, Heirtzler D, Izmodenov V V, Kubiak M, Kucharek H, Lee M A, Leonard T, McComas D J, Petersen L, Saul L, Scheer J A, Schwadron N, Witte M and Wurz P 2009 *Science* **326** 969–

[8] Krimigis S M, Mitchell D G, Roelof E C, Hsieh K C and McComas D J 2009 *Science* **326** 971–

[9] Schwadron N A, Adams F C, Christian E R, Desiati P, Frisch P, Funsten H O, Jokipii J R, McComas D J, Moebius E and Zank G P 2014 *Science* **343** 988

[10] Compton A H and Getting I A 1935 *Physical Review* **47** 817–821

[11] Gleeson L J and Axford W I 1968 *Astrophys. J.* **154** 1011

[12] The Tibet Asγ Collaboration, Amenomori M, Bi X J, Chen D, Cui S W, Danzengluobu, Ding L K, Ding X H, Fan C, Feng C F, Feng Z, Feng Z Y, Gao X Y, Geng Q X, Guo H W, He H H, He M, Hibino K, Hotta N, Hu H, Hu H B, Huang J, Huang Q, Jia H Y, Kajino F, Kasahara K, Katayose Y, Kato C, Kawata K, Labaciren, Le G M, Li A F, Li J Y, Lou Y Q, Lu H, Lu S L, Meng X R, Mizutani K, Mu J, Munakata K, Nagai A, Nanjo H, Nishizawa M, Ohnishi M, Ohta I, Onuma H, Ouchi T, Ozawa S, Ren J R, Saito T, Saito T Y, Sakata M, Sako T K, Shibata M, Shiomi A, Shirai T, Sugimoto H, Takita M, Tan Y H, Tateyama N, Torii S, Tsuchiya H, Udo S, Wang B, Wang H, Wang X, Wang Y, Wang Y G, Wu H R, Xue L, Yamamoto Y, Yan C T, Yang X C, Yasue S, Ye Z H, Yu G C, Yuan A F, Yuda T, Zhang H M, Zhang J L, Zhang N J, Zhang X Y, Zhang Y, Yizhang, Zhaxisangzhu, Zhou X X and The Tibet ASγ Collaboration 2011 *Advances in Space Research* **47** 629–639

[13] Abbasi R, Abdou Y, Abu-Zayyad T, Adams J, Aguilar J A, Ahlers M, Andeen K, Auffenberg J, Bai X, Baker M and et al 2010 *Astrophys. J. Lett.* **718** L194–L198 (*Preprint* `1005.2960`)

[14] Abbasi R, Abdou Y, Abu-Zayyad T, Ackermann M, Adams J, Aguilar J A, Ahlers M, Allen M M, Altmann D, Andeen K and et al 2012 *Astrophys. J.* **746** 33 (*Preprint* `1109.1017`)

[15] Abdo A A, Allen B, Aune T, Berley D, Blaufuss E, Casanova S, Chen C, Dingus B L, Ellsworth R W, Fleysher L, Fleysher R, Gonzalez M M, Goodman J A, Hoffman C M, Hüntemeyer P H, Kolterman B E, Lansdell C P, Linnemann J T, McEnery J E, Mincer A I, Nemethy P, Noyes D, Pretz J, Ryan J M, Parkinson P M S, Shoup A, Sinnis G, Smith A J, Sullivan G W, Vasileiou V, Walker G P, Williams D A and Yodh G B 2008 *Physical Review Letters* **101** 221101 (*Preprint* `0801.3827`)

[16] Bzowski M, Kubiak M A, Moebius E, Bochsler P, Leonard T, Heirtzler D, Kucharek H, Sokol J M, Hlond M, Crew G B, Schwadron N A, Fuselier S and McComas D J 2012 *Astrophys. J. Suppl.* **In Press**

[17] Amenomori M, Bi X J, Chen D, Cui S W, Danzengluobu, Ding L K, Ding X H, Fan C, Feng C F, Feng Z, Feng Z Y, Gao X Y, Geng Q X, Guo H W, He H H, He M, Hibino K, Hotta N, Hu H, Hu H B, Huang J, Huang Q, Jia H Y, Kajino F, Kasahara K, Katayose Y, Kato C, Kawata K, Labaciren, Le G M, Li A F, Li J Y, Lou Y Q, Lu H, Lu S L, Meng X R, Mizutani K, Mu J, Munakata K, Nagai A, Nanjo H, Nishizawa M, Ohnishi M, Ohta I, Onuma H, Ouchi T, Ozawa S, Ren J R, Saito T, Saito T Y, Sakata M, Sako T K, Shibata M, Shiomi A, Shirai T, Sugimoto H, Takita M, Tan Y H, Tateyama N, Torii S, Tsuchiya H, Udo S, Wang B, Wang H, Wang X, Wang Y, Wang Y G, Wu H R, Xue L, Yamamoto

Y, Yan C T, Yang X C, Yasue S, Ye Z H, Yu G C, Yuan A F, Yuda T, Zhang H M, Zhang J L, Zhang N J, Zhang X Y, Zhang Y, Yizhang, Zhaxisangzhu, Zhou X X and The Tibet ASγ Collaboration 2011 *Proc. of 32nd ICRC* (Beijing, China) p paper 0379

[18] Frisch P C, Redfield S and Slavin J D 2011 *Ann. Rev. Astron. Astrophys.* **49** 237–279

[19] Frisch P C, Grodnicki L and Welty D E 2002 *Astrophys. J.* **574** 834–846

[20] Wood B E, Linsky J L and Zank G P 2000 *Astrophys. J.* **537** 304–311

[21] Frisch P C and Mueller H R 2011 *Space Sci. Rev.* 130 (*Preprint* 1010.4507)

[22] Lallement R, Bertin P, Ferlet R, Vidal-Madjar A and Bertaux J L 1994 *Astron. Astrophys.* **286** 898–908

[23] Hébrard G, Mallouris C, Ferlet R, Koester D, Lemoine M, Vidal-Madjar A and York D 1999 *Astron. Astrophys.* **350** 643–658 (*Preprint* arXiv:astro-ph/9909061)

[24] Linsky J L, Rickett B J and Redfield S 2008 *Astrophys. J.* **675** 413–419

[25] Redfield S and Linsky J L 2008 *Astrophys. J.* **673** 283–314 (*Preprint* arXiv:0709.4480)

[26] Smirnova T V, Gwinn C R and Shishov V I 2006 *Astron. Astrophys.* **453** 601–607 (*Preprint* arXiv:astro-ph/0603490)

[27] Bzowski M, Kubiak M A, Möbius E, Bochsler P, Leonard T, Heirtzler D, Kucharek H, Sokół J M, Hłond M, Crew G B, Schwadron N A, Fuselier S A and McComas D J 2012 *Astrophys. J. Suppl.* **198** 12 (*Preprint* 1202.0415)

[28] Möbius E, Bochsler P, Bzowski M, Heirtzler D, Kubiak M A, Kucharek H, Lee M A, Leonard T, Schwadron N A, Wu X, Fuselier S A, Crew G, McComas D J, Petersen L, Saul L, Valovcin D, Vanderspek R and Wurz P 2012 *Astrophys. J. Suppl.* **198** 11

[29] Wolleben M 2007 *Astrophys. J.* **664** 349

[30] Schwadron N A, Allegrini F, Bzowski M, Christian E R, Crew G B, Dayeh M, DeMajistre R, Frisch P, Funsten H O, Fuselier S A, Goodrich K, Gruntman M, Janzen P, Kucharek H, Livadiotis G, McComas D J, Moebius E, Prested C, Reisenfeld D, Reno M, Roelof E, Siegel J and Vanderspek R 2011 *Astrophys. J.* **731** 56

[31] Schönrich R, Binney J and Dehnen W 2010 *MNRS* **403** 1829–1833 (*Preprint* 0912.3693)

[32] Shalchi A, Büsching I, Lazarian A and Schlickeiser R 2010 *Astrophys. J.* **725** 2117–2127

Journal of Physics: Conference Series **531** (2014) 012011 doi:10.1088/1742-6596/531/1/012011

Heliospheric Boundary and the TeV Cosmic Ray Anisotropy

Paolo Desiati[1,2], **Alex Lazarian**[2]

1. Wisconsin IceCube Particle Astrophysics Center (WIPAC), University of Wisconsin, Madison, WI 53703
2. Department of Astronomy, University of Wisconsin, Madison, WI 53706

E-mail: desiati@wipac.wisc.edu

Abstract. Observations over the last few decades have shown that cosmic rays have a small non uniform distribution in arrival direction. Such anisotropy appears to have a roughly consistent topology between 10's GeV and 100's TeV, with a smooth energy dependency on phase and amplitude. However, above a few 100's TeV a sudden change in the topology of the anisotropy is observed. The cosmic ray arrival directions are expected to depend on the distribution of their sources in the Milky Way, as well as on effects arising from propagation in the inhomogeneous and turbulent interstellar magnetic field. In particular, in the 1-10 TeV energy range, the gyroradius of cosmic ray particles, much smaller that the injection scale of interstellar magnetic field turbulence, is comparable to the size of the heliosphere. Resonant scattering processes of cosmic ray particles propagating through the heliosphere may be able to induce a redistribution of the small anisotropic component of their flux. In this paper we discuss on the processes that occurs to TeV cosmic rays in the heliosphere and on the possibility to probe the heliospheric magnetic fields on large scale by studying the arrival distribution of cosmos rays.

1. Introduction

In the last few decades a number of experiments have provided long term and high statistical significance evidence of a global anisotropy in the cosmic ray arrival distribution with a relative amplitude of order $10^{-4} - 10^{-3}$. Observations in the northern hemisphere were reported from energies of 10's to several 100's GeV with muon detectors [1, 2, 3], in the 1-10's TeV energy range with various surface arrays and underground detectors [4, 5, 6, 7, 8, 9, 10, 11, 12, 13, 14], up to 100's TeV [15]. Recently, similar observations of a large angular scale anisotropy were reported in the southern hemisphere by the IceCube Observatory, in the energy range between 10 TeV and a few PeV [16, 17, 18]. These results show that the anisotropy matches the observations in the northern hemisphere in the 10 TeV energy scale (see top panel of figure 1), but that it changes topology at energy in excess of about 100 TeV, in a fashion that may be interpreted as a directional flip, confirming the observations by [15]. The amplitude of the high energy anisotropy, characterized by a wide deficit on a relatively uniform flux, increases with energy [18]. The transition between the two anisotropy regimes is preceded by a steady decrease of the amplitude with energy above 10 TeV, following an increase trend at lower energies (see [4] for instance). This might suggest that the 10 TeV energy scale is where the transition starts with the interference between the two different distributions disrupting the energy dependence

Journal of Physics: Conference Series **531** (2014) 012011 doi:10.1088/1742-6596/531/1/012011

of the anisotropy amplitude.

The global anisotropy cannot be described with a simple dipole, but as a superposition of spherical harmonic contributions, where statistically significant small angular scale features (i.e. less than 30°), with amplitude of order $10^{-5} - 10^{-4}$, are also observed [19, 14], in agreement with similar observations in the northern hemisphere [20, 21, 22, 13] (see the bottom panel of figure 1).

The origin of the observed anisotropy is not understood yet, however it is reasonable to assume that it is a combination of effects correlated to the distribution of the galactic sources of cosmic rays, of the geometry and turbulent properties of the galactic magnetic field and on propagation in this field. These are likely also the responsible of the complex shape of the energy spectrum [23]. Since we don't know where the sources of cosmic rays are and we don't know the details of the interstellar magnetic field, understanding the observations is not an easy task.

The large scale anisotropy can be qualitatively explained on the basis of diffusive propagation of cosmic rays in the Milky Way from stochastically distributed sources. Numerical studies show that it is possible to find a given realization of galactic source distribution that might explain the observed non monotonic energy dependence of the anisotropy amplitude, although for an ensemble of realizations the mean amplitude is overestimated compared to observations [24, 25, 26, 27, 28]. Even accounting for a more likely anisotropic diffusion, observations appear to be outliers compared to predictions [29, 30]. The fact that cosmic ray anisotropy is not a simple dipole but can be mostly explained with a superposition of a dipole and quadrupole term, seems to suggest that other transport processes might be important as well. For instance drift diffusion driven by a gradient of cosmic ray density in the local interstellar medium, producing a bi-directional anisotropy, was considered by [20, 31].

The small scale anisotropy may be produced by the interactions of cosmic rays with isotropically turbulent interstellar magnetic field. Scattering processes with stochastic magnetic instabilities produce perturbations in the arrival direction distribution of an anisotropic distribution of cosmic ray particles within the scattering mean free path. Such perturbations may be observed as stochastic localized excess or deficit regions [32, 33], and the corresponding angular power spectrum can be analytically predicted from the Liouville Theorem [34]. The injection scale of interstellar turbulence is of order 10 pc within the galactic arms and 100 pc in the inter arm regions [35]. In the cascading processes down to smaller scales, the turbulent eddies become elongated along the magnetic field lines [36, 37]. This anisotropic turbulence makes scattering processes inefficient, and scattering mean free path can be larger than turbulence injection scale, so that particles basically stream along magnetic field lines with small cross fieldline transport [38, 39].

Besides the cascading interstellar magnetic field turbulence down to damping scale (typically of order 0.1 pc), there are other sources of magnetic perturbations on smaller scales. The closest to Earth is represented by the heliosphere, formed by the interaction between the solar wind and the interstellar flow, about 600 AU wide and extending several thousands astronomical units downstream the interstellar wind [40]. Globally, the heliosphere constitutes a perturbation in the 3 μG Local Interstellar Magnetic Field (LIMF) with injection scale comparable to $\sim$10 TeV proton gyroradius. It is therefore reasonable that the LIMF draping around the heliosphere might be a significant source of resonant scattering, capable of redistributing the arrival directions of TeV cosmic cosmic ray particles.

In this paper we briefly summarize the current understanding of the effects of the heliosphere on TeV cosmic rays and whether we can use the anisotropy observations as diagnostics on the heliospheric boundary with the local interstellar medium.

Journal of Physics: Conference Series **531** (2014) 012011 doi:10.1088/1742-6596/531/1/012011

2. The Heliosphere

The solar system is located approximately 8 kpc from the galactic center and it is dragged by the global galactic rotation, similarly to the Local Interstellar Medium (LISM). However, the relative motion of the solar system with respect to the LISM is determined by the expansion of the Loop I superbubble from the Scorpion-Centaurus Association (at a distance of about 60 pc), which is near the apparent direction of the galactic center [41]. The solar system is currently located within an expanding shell of Loop I, which is breaking up in smaller cloudlets that share similar composition and velocities. And the LIMF appears to be almost parallel to the shell front and relatively coherent within a distance of 80-100 pc. One of these cloudlets, the so called Local Interstellar Cloud (LIC) is where the solar system is located, with a relative velocity to the LIC of approximately 23 km/sec [42]. The interaction between the supersonic outward solar wind and the incoming interstellar flow produces a comet shaped interface where the interplanetary and the interstellar magnetic fields rearrange. At a distance of about 80 AU from Earth the termination shock is formed where the solar wind becomes subsonic, but the separation between the heliospheric and the interstellar medium occurs at a distance of about 120 AU in the upstream direction, where the heliopause is located. From numerical simulations, the heliosphere is expected to be as wide as 600 AU and to form an elongated tail in the downstream direction, that can be long several 1000's AU [40].

Since the LISM is a partially ionized medium, neutral particles can penetrate through the heliopause and have charge exchange with the highly ionized solar wind. Such processes create instabilities on the heliopause in the upstream direction, that propagate along the flanks and that can reach spatial scales of 10-100 AU [43]. Another source of instabilities on the heliopause is represented by the solar cycles, which affect the heliospheric magnetic field pressure, that regulates the distance of the heliopause from Earth [44]. The magnetic field polarity inversions due to solar cycles, produce unipolar regions of magnetic fields that, dragged by the solar wind, drift along the heliotail [40].

2.1. Heliospheric Influence on TeV Cosmic Rays

Globally, the heliosphere is a perturbation of the LIMF on a scale of about 600 AU. In addition, complex structures in the inner heliospheric magnetic field of about 100-200 AU and instabilities on the heliopause on scales of approximately 10-100 AU, have necessarily an effect on the charged particles that propagate through.

In [45] the scenario of TeV cosmic ray interaction with the heliosphere was first presented as a possible cause of the observed anisotropy. It is known that galactic cosmic rays below 100 GeV are strongly influenced by solar activity and by the geometry and small scale instabilities (sub-AU scale) of the interplanetary magnetic field within the termination shock. At TeV energies such small scale effects are subdominant because the gyroradius of a proton is larger than the termination shock size, in the inner heliospheric magnetic field. However TeV cosmic rays are in gyro-resonance with the large scale heliospheric magnetic field structure, with unipolar magnetic regions in the inner heliotail and with the instabilities on the boundary with the LISM, depending on the local pitch angle. Because of this, the heliosphere can actually have a significant impact on the distribution of TeV cosmic rays that cross the heliosphere.

Figure 1 shows the sky map of the relative intensity of the global cosmic ray anisotropy (on the top) and the sky map of statistical significance of the small scale anisotropy (on the bottom) obtained by filtering out all the cosmic ray count gradients compatible with a dipole and a quadrupole dependence. In both maps the observations from IceCube, at median particle energy of 20 TeV, are shown. In the northern hemisphere the figures show the results from Tibet-ASγ (with median energy of 5 TeV, on the top panel of the figure) and from Milagro (with median energy of about 1 TeV, on the bottom panel of the figure). The figures show a degree of matching compatibility between the different observations from the northern and

Journal of Physics: Conference Series **531** (2014) 012011 doi:10.1088/1742-6596/531/1/012011

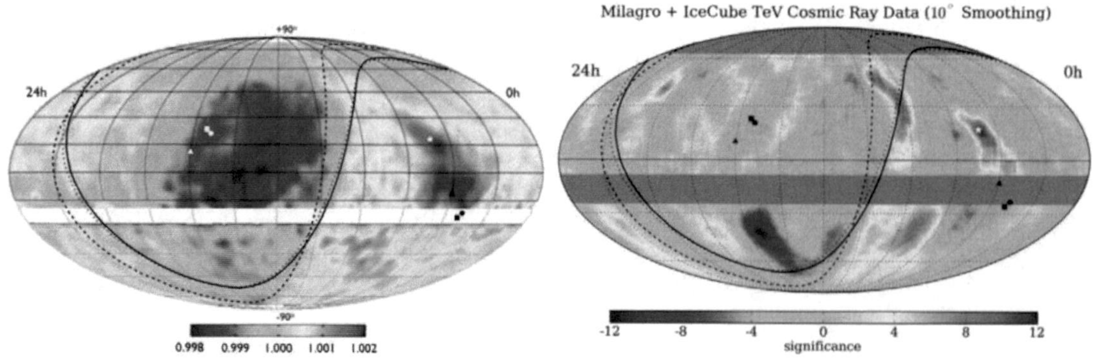

Figure 1. *Top*: map in equatorial coordinates constructed by combining relative intensity distributions of cosmic ray counts independently normalized within declination bands of order $1° - 5°$. The map shows the observation by Tibet ASγ at about 5 TeV in the northern hemisphere [6] and by IceCube at a median energy of 20 TeV in the southern hemisphere [16]. *Bottom*: map in equatorial coordinates of the statistical significance of the observed cosmic ray counts. The map shows the observation by Milagro at about 1 TeV in the northern hemisphere (from [21]) and that by the IceCube Observatory at a median energy of 20 TeV in the southern hemisphere (from [19]). Also shown the equator "at infinity" (see text) in black continuous line, corresponding to the direction of LIMF inferred in [46, circular symbols], that in black dotted line from [47, square symbols], and that in black dashed line from [48, triangle symbols]. The star symbol indicates the downwind interstellar medium flow with coordinate (5 hr, +17°). Color scale indicates relative excess (red) and deficit (blue) with respect to average intensity. From [45].

southern hemispheres, the differences possibly attributed to the different sensitivities of the cited experiments on the energy and, possibly, mass composition of the cosmic rays. Some visible features seem ordered by the LIMF (with three direction estimations represented with circle, square and triangle symbols in figure 1), in particular the fact that the magnetic equator "at infinity" (i.e. the plane perpendicular to the unperturbed LIMF and passing by Earth, and represented with black lines in figure 1) seems to be aligned with the boundary region between the excess and the deficit of the anisotropy. But there appears to be some ordering by the heliotail direction as well (almost coinciding with the downstream interstellar medium flow direction, represented with a white star in figure 1).

At 10 TeV, cosmic ray protons streaming along the LIMF have gyroradii comparable to the heliospheric size, thus making it possible for them to undergo resonant scattering. Therefore, 10 TeV represents a transition scale between the lower energies, where cosmic rays follows adiabatically the magnetic field lines, and the higher energies, where the heliospheric influence starts to become subdominant, making it possible for the interstellar-originated anisotropy to grow progressively visible. This is in general agreement with the energy dependence of the anisotropy amplitude that increases up to about 10 TeV and decreases above this energy [4]. Resonant scattering may be able to redistribute the 10^{-3} interstellar anisotropic component of the cosmic ray arrival distribution, depending on particle energy and pitch angle, and on the properties of heliospheric magnetic field inside the tail and on the heliopause. Backscattering on the magnetic instabilities at the boundary with the LISM (see figure 2 for a simple illustration) may have the effect to locally amplify particle density gradients which can be interpreted as localized features [45].

The localized excess region of cosmic rays observed in the direction of the heliotail, highly pronounced in the sub-TeV energy range [1], seems to be characterized by a harder than average energy spectrum up to about 10 TeV [21, 22]. If confirmed, such observations might indicate that

Journal of Physics: Conference Series **531** (2014) 012011 doi:10.1088/1742-6596/531/1/012011

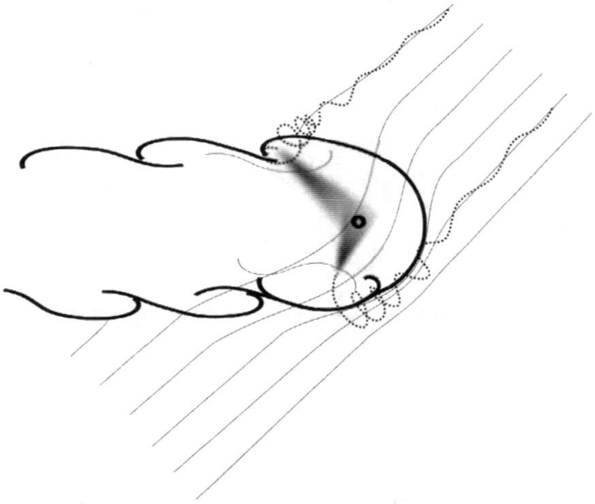

Figure 2. Illustration of the turbulent ripples at large scale on the outer heliospheric boundary. High energy cosmic rays propagating from the upstream direction into the front of the heliosphere resonantly scatter with the trailing magnetic field perturbations on the flanks of the heliosphere and along the heliotail. The uniform pitch angle distribution makes it possible for particles to scatter back to Earth. The grey blurred area schematically represents the possible scattered particle trajectories toward Earth. From [45].

the small fraction (10^{-4}) of cosmic rays from this narrow angular region might be reaccelerated in magnetic reconnection processes occurring in between the unipolar magnetic domains generated by solar cycles [49, 50]. Such reacceleration processes would only involve a negligible fraction of the total turbulent energy transported by the solar wind across the heliotail.

In [51, 52] numerical calculations show that TeV cosmic ray arrival direction distribution, originally ordered by the LIMF, is perturbed so that they appear to be ordered by their entry into the heliosphere, assuming the direction of the interstellar medium flow velocity and of the LIMF as evaluated from the analysis of data by the Interstellar Boundary Explorer (IBEX). If a dipole anisotropy is assumed, either from the contribution of a nearby recent source of cosmic rays or simply from the convective gradient due to the relative motion with respect to the LISM, the deformation induced by the LIMF draping around the heliosphere was estimated and found to globally resemble the actual observations.

Another heliospheric scenario appeals to the possibility that electric fields arising from the motion of the heliospheric plasma might be able to produce the observed localized excess regions of TeV cosmic rays [53].

3. Conslusion

The possibility that the heliosphere may actually shape the high energy cosmic rays arrival distribution is reasonable based on purely dimensional arguments. State of the art numerical simulations of the heliosphere, that include dynamical processes such as charge exchange between neutral and ionized components of the medium, solar cycles and plasma instabilities are starting to provide a clearer idea of the properties of the heliosphere at large distances from Earth and, at the same time, reproduce observations by spacecrafts [40, 44]. The study of how such properties may influence TeV cosmic ray arrival direction distribution will provide the basis of exploring the boundary region between the heliosphere and the LISM with cosmic rays, and it will open the doors to the understanding of our local interstellar medium, as well. Numerical studies currently under development will be reported in a future publication.

References

[1] Nagashima et al. 1998 *J. of Geophys. Res.* **1031** 17429.
[2] Hall, D. L. et al. 1999 *J. of Geophys. Res.* **104** 6737.
[3] Munakata, K. et al. 2010 *Astrophys. J.* **712** 1100.
[4] Amenomori, M., et al. 2005 *Astrophys. J. Lett.* **626** L29.
[5] Amenomori, M., et al. 2006 *Science* **314** 439.
[6] Amenomori, M., et al. 2011 *Proc. 32nd ICRC* Beijing China.
[7] Guillian, G., et al. 2007 *Phys. Rev.* D **75** 062003.
[8] Abdo, A. A., et al. 2009 *Astrophys. J.* **698** 2121.
[9] Zhang, J. L. 2009 *Proc. 31st ICRC* Lódź, Poland.
[10] Shuwang, C.: 2011 *Proc. 32nd ICRC* Beijing China.
[11] Di Sciascio, G. 2013 *EPJ Web Conf.* **52** 04 004.
[12] de Jong, J. 2011 *Proc. 32nd ICRC* Beijing, China.
[13] BenZvi, S. Y. et al. 2013 *Proc. 33rd ICRC* Rio de Janeiro, Brazil.
[14] Santander, M. et al. 2013 *Proc. 33rd ICRC* Rio de Janeiro, Brazil.
[15] Aglietta, M., et al. 2009 *Astrophys. J. Lett.* **692** L130.
[16] Abbasi, R., et al. 2010 *Astrophys. J. Lett.* **718** L194.
[17] Abbasi, R., et al. 2012 *Astrophys. J.* **746** 33.
[18] Aartsen, M. et al. 2013 *Astrophys. J.* **765** 55.
[19] Abbasi, R., et al. 2011 *Astrophys. J.* **740** 16.
[20] Amenomori, M., et al. 2007 *Proc. 30th ICRC* Mérida, Mexico.
[21] Abdo, A. A., et al. 2008 *Phys. Rev. Lett.* **101** 221 101.
[22] Bartoli, B., et al. 2013 *Phys. Rev.* D **88-8** 082001.
[23] Gaisser, T. K. et al. 2013 *Front. of Phys.* **8-6** 748.
[24] Erlykin A. D., & Wolfendale A. W. 2006 *Astropart. Phys.* **25** 183.
[25] Blasi, P., & Amato, E.: 2012 *JCAP* **1** 11.
[26] Ptuskin V. 2012 *Astropart. Phys.* **39** 44.
[27] Pohl, M. & Eichler, D. 2013 *Astrophys. J.* **766** 9.
[28] Sveshnikova, L. G. et al. 2013 *Astropart. Phys.* **50** 33.
[29] Effenberger, F. et al. 2012 *A&A* **547** A120.
[30] Kumar R. & Eichler D. 2014 *Astrophys. J.* **785** 129.
[31] Mizoguchi Y. et al. 2009 *Proc. 31st ICRC* Łódź, Poland.
[32] Giacinti G. & Sigl G. 2012 *Phys. Rev. Lett.* **109** 071101.
[33] Biermann P. L. et al. 2013 *Astrophys. J.* **768** 124.
[34] Ahlers M. 2014 *Phys. Rev. Lett.* **112** 021101.
[35] Haverkorn, M. et al. 2006 *Astrophys. J. Lett.* **637** L33.
[36] Sridhar S. & Goldreich P. 1994 *Astrophys. J.* **432**, 612.
[37] Goldreich P. & Sridhar S. 1995 *Astrophys. J.* **438**,763.
[38] Yan H. & Lazarian A. 2008 *Astrophys. J.* **673** 942.
[39] Lazarian A. & Yan H. 2014 *Astrophys. J.* **784** 38.
[40] Pogorelov N. V. et al. 2009 *Astrophys. J.* **696** 1478.
[41] Frisch P. C. et al. 2011 *Annu. Rev. Astron. Astrophys.* **49** 237.
[42] McComas D. J. et al. 2012 *Science* **336** 1291.
[43] Borovikov S. N. et al. 2008 *Astrophys. J.* **682** 1404.
[44] Pogorelov N. V. et al. 2013 *Astrophys. J.* **772** 2.
[45] Desiati P. & Lazarian A. 2013 *Astrophys. J.* **762** 44.
[46] Funsten, H. O., et al. 2009 *Science* **326** 964.
[47] Schwadron N. A., et al. 2009 *Science* **326** 966.
[48] Heerikhuisen, J., et al. 2010 *Astrophys. J. Lett.* **708** L126.
[49] Lazarian A. & Desiati P. 2010 *Astrophys. J.* **722** 188.
[50] Desiati P. & Lazarian A. 2012 *NPG* **19** 351.
[51] Schwadron N. et al. 2014 *Science* **343** 988.
[52] Schwadron N. et al. 2014, *in these proceedings.*
[53] Drury L. O'C. 2013 *Proc. 33rd ICRC* Rio de Janeiro, Brazil.

AUTHOR INDEX